A World of Looms
Weaving Technology and Textile Arts

神机妙算
世界织机与织造艺术

Chief Editors
Zhao Feng, Sandra Sardjono, Christopher Buckley

Foreword

According to Joseph Needham, the character 机 (*ji*) in ancient Chinese has two meanings:
(1) a loom
(2) a matter of wit and intelligence

These definitions equate weaving on a loom with the ability to learn and invent. People from all corners of the world share knowledge of looms and weaving skills; yet, different regions in different periods develop specific looms to produce unique textile types. Thus, it is appropriate that an exhibition on looms at the China National Silk Museum (NSM), Hangzhou, should be called *Shenji Miaosuan* (神 机 妙 算), which literally translates as "celestial looms and marvelous calculations."

This book is based on this exhibition, titled in English *A World of Looms: Weaving Technology and Textile Arts in China and Beyond* (hereinafter "*A World of Looms*"), held from 30 May to 15 September 2018. It was the first exhibition in China to present the rich cultural heritage of looms and weaving technologies from around the world. More than 50 looms and their related textiles were selected for display. These objects celebrated the march of textile innovation through the lens of global textile traditions, contextualized into their spatial and temporal distributions.

The exhibit began with the China section, which showcased loom models from recent archaeological excavations in the Silk Road Gallery and classical Chinese looms and looms of Chinese ethnic minorities in the Weaving Gallery. These permanent galleries highlighted the vital role of cultural exchanges along the Silk Road for the evolution of loom types in China and its bordering regions. Next were the temporary installations. The Eurasia section mapped the various mechanisms and patterning systems amongst looms in different parts of Asia and Europe; the Americas section featured body-tensioned looms of the Peruvians; the Africa section displayed traditional looms from Ghana. Ending the timeline explored in the exhibition was the Jacquard loom as well as digital Jacquard looms of the type that are now the mainstay of commercial weaving around the world. All the Chinese looms were drawn from the museum's own collection, but most of the others were newly acquired for the purpose of the exhibition.

The goals for the exhibition *A World of Looms* were multi-fold. First, we aimed to highlight the similarities and differences among loom types from different regions. Second, we hoped to raise questions about loom technological evolution in regard to the mechanics and patterning systems. Third, we sought to explore some of the various vectors of loom technology transmission around the world. Many looms were chosen for the exhibition because they represent unique technological solutions, ranging from the simple body-tensioned looms to drawlooms and Jacquard looms.

In conjunction with the exhibition an international conference was convened on 31 May 2018, bringing together 13 textile scholars specializing in different regions. They were invited to give lectures about looms within their areas of study. The topics covered included mechanical structures, the origins of looms, and traditional weaving practices. One of the major outcomes of the conference was a confirmation that the study of looms benefited greatly from a wider, comparative perspective.

Various loom workshops were also held from 1–3 June. Weavers were invited to operate the looms that were on display and demonstrate

Participants of the International Conference and Workshops

their weaving. Some specialists delivered lectures on their researches on looms. These live demonstrations were extremely important because looms as static museum objects cannot fully capture weaving processes; the act of collecting itself stops the activity on the loom. For example, a collected loom may not record all the patterning systems that were used in creating a given textile, as some of the systems were only temporarily fitted on the loom during weaving. Seeing how weavers operate the looms' mechanics and patterning systems completes our learning experience and brings these looms to life.

We hope that this book captures the lively and collaborative spirit of the exhibition *A World of Looms*, conference, and workshops. Given the natural limitations on scope, the catalog part offers representative types, not a comprehensive guide to all the looms around the world. The essays are authored by many of the scholars who spoke at the conference and by a few new contributors. Each essay reflects its author's individual viewpoint. Many of the weavers who conducted the workshops also contribute to this book through recorded interviews, which present their opinions about the art of weaving within their own traditions. The exhibition *A World of Looms* is one of a kind because of the involvement of these weavers, from whom we learned so much. It is our belief that we need to continue to learn from them because the stories of looms can never be complete without the voices of the weavers.

Zhao Feng, Curator and Director
China National Silk Museum, Hangzhou

Sandra Sardjono, Co-Curator
Tracing Patterns Foundation, California

Long Bo, Assistant Curator
China National Silk Museum, Hangzhou

Acknowledgments

The exhibition *A World of Looms* and its associated conference and workshops are the result of the collaborative efforts of institutions and individuals from around the world. The China National Silk Museum conceived the idea of the exhibition in mid-2017, after which the project received an outpouring of support from many institutions and individuals. It is impossible to list all the contributions we have received in full, and we apologize in advance for any omissions.

Our sincere gratitude goes first and foremost to the conference and workshop participants who have lent their expertise and donated much time in shaping the exhibition content and enriching the overall program. Their names are listed here in alphabetical order: Hemang Agrawal, Naseem Ahmad, Eva Andersson Strand, Tauseef Ahmad Ansari, Zainal Arifin, Bang Yeonok, Bernhard Bart, Eric Boudot, Christopher Buckley, Flora Callanaupa, Carol Cassidy, Erika Dubler, Andrée Etheve, Ho Zhaohua, Ata Jallayer, Malika Kraamer, Luo Qun, Rasuljon Mirzaahmedov, Rastjouy Ardakani Mirzamohammad, Binafsha Nodir, Magdalena Öhrman, Bouakham Phengmixay, Elena Phipps, Vankar Babubhai Ratanshi, Sandra Sardjono, Guy Scherrer, Sim Yeonok, Yanet Soto, Nicolas Stoll, Tatyana Trudolyubova, Vankar Vishram Valji, Vankar Rajesh Vishram, Gillian Vogelsang-Eastwood, Yoshimoto Shinobu, Yuma Taru, and Yu Youde.

Many others contributed their knowledge, collections, and precious time in the preparation of the exhibition and this book. We would like to express our appreciation to Ishii Shinobu of the Miho Museum, Nakajima Yoichi of the Nakajima Yoichi Silk Weaving Studio, Kajitani Nobuko, Mary Connors, Geneviève Duggan, Sandra Niessen, Ompu Elza, Anita Gathmiri, Ni Gede Diar, Ngurah Hendrawan, Ida Ayu Puniari, Dinny Jusuf, I Putu Cobby Wiryadi, Assadour Markarov, An Weizhu, Judith Cameron, and Linda Susan McIntosh.

We are grateful to the many individuals and institutions who generously hosted and assisted the museum researchers during their travels to seek looms, textiles, and information. In Thailand, we received assistance from Sarttarat Muddin of Queen Sirikit Textile Museum, Prang Rojanachotikul of Jai-Sook Studio, and Patricia Cheesman of Naenna Studio. We also appreciate the help of Pav Eang Khoing of Angkor Craftsman Silk Company in Cambodia and the gift of an ironwood loom from Vu Cao Trung of the Vietnam Sericulture Association. In Indonesia, we were warmly received by Mis Ari of the Textile Museum Jakarta, Intan Mardiana of the National Museum Indonesia, G.K.B.R.A.A. Paku Alam of the Ndalem Pakualaman Palace in Yogyakarta, Anggi Bambang of the Traditional Textile Arts Society of Southeast Asia, Herman and Eka Rachman of the Pesona Bari Songket, Haji Imam of Tenun Imam, and William Ingram and Jean Howe of the Threads of Life. We especially want to thank Siti Mariah Waworuntu for her hospitality and insightful counsel. In India, we were given invaluable advice by Vijay Ramchandani and Neel Kamal Chapagain of Ahmedabad University, Rahul Jain of The Calico Museum of Textiles, and people from the Gujarat Weavers' Service Center. In Egypt, we received thoughtful guidance from Saied Hamed of the Egypt Archaeological Museum. In Israel, we were assisted by Yonit Crystal to acquire a Beduin loom. In France, Michel Rodarie and Xavier de la Selle of Musée Gadagne of Lyon helped us to obtain a Jacquard loom. We also received exceptional support in China from Li Mingbin, Zhou Xun, and Zhang Baolin from Chengdu Museum; Wang Yi, Xiao Lin, Xie

Tao, Yang Tao, and Li Yang from the Chengdu Institute of Cultural Relics and Archaeology; Wu Weifeng, Tang Jianlin, and Li Xia from the Museum of Guangxi Zhuang Autonomous Region; Ma Wendou, Fan Haitao, and Ping Li from the Yunnan Provincial Museum; Liu Xinyu from Jing'an Museum; Liu Shuangshuang from the International Silk Union; Wang Ziqiang, Wang Shujuan, Yang Hailiang, Dai Huali, Xu Qingqing, and Ye Ye from NSM; Wei Wenquan and Mao Yadong from PT (Zhejiang) Machinery of Textile Co. Ltd.; Zhu Wei and Zhu Hexing from Haiyan Weicheng Print Co. Ltd.; Lu Jianping and Jin Wei from Haining Tianyi Textile Co. Ltd.; Qian Zhiyi, Li Kun, and Ma Zhenjun from Zhejiang Huiming Jacquard Weaving Co. Ltd.; and Miao Yuhen from Hangzhou Dujinsheng Industrial Co. Ltd.

On a personal note, I would like to thank my co-curator, Sandra Sardjono, who worked closely with me starting from the inception of this exhibition and throughout its making. Her guidance in my field research in Indonesia, her assistance in the organization of the conference and workshops, and her co-editing of this book were invaluable. Editing a multi-authored volume on technical subjects such as looms is a big challenge, and I greatly appreciate the diligent work of my other co-editor Christopher Buckley.

Last and most importantly, I would like to acknowledge the incredible achievements of the NSM research, curatorial, and exhibition teams, headed by Long Bo, the assistant curator. He, Irene Lu, and Lu Jialiang were the backbone of the overall project of *A World of Looms*. The beautiful design and installation of the exhibit were done by Zhang Yi, Zhou Yang, Wang Ziqiang, Shen Guoqing, Zhao Fan, Sophia Liang, and the conservation team. Without their support the project *A World of Looms* would not have materialized. Thank you all!

Zhao Feng, Director
China National Silk Museum, Hangzhou

Notes

Looms and textiles depicted in this book belong to the China National Silk Museum, unless stated otherwise.

List of Contributors

Hemang AGRAWAL (HA), Creative Director, The Surekha Group and Label-Hemang Agrawal, Varanasi, India

Naseem AHMAD (NA), Weaver, Varanasi, India

Eva ANDERSSON STRAND (EAS), Associate Professor and Director, Centre for Textile Research, University of Copenhagen, Denmark

Tauseef Ahmad ANSARI (TAA), Weaver, Varanasi, India

Zainal ARIFIN (ZA), Director, Zainal Songket, Palembang, Indonesia

Bernhard BART (BB), Director, Studio Songket, Minangkabau, Indonesia

Eric BOUDOT (EB), Independent Researcher, Beijing, China

Christopher BUCKLEY (CB), Independent Researcher, Oxford, UK

Flora CALLANAUPA (FC), Weaver, Centro de Textiles Tradicionales, Cusco, Peru

Judith CAMERON (JC), Professor, The Australian National University, Canberra, Australia

Carol CASSIDY (CC), Director, Lao Textiles, Vientiane, Laos

Bob DENNIS (BD), Weaver, Tema, Ghana

Andrée ETHEVE (AE), Director, Femmes Entreprenneurs Environnement Mahajanga, Madagascar

HO Zhaohua (HZ), Professor, Fu Jen Catholic University, Taiwan, China

Ata JALLAYER (AJ), PhD Candidate, University of Science and Technology of China, Hefei, China

Dinny JUSUF (DJ), Director, Torajamelo, Indonesia

Malika KRAAMER (MK), Independent Researcher, Leicester, UK

LONG Bo (LB), Assistant Curator, China National Silk Museum, Hangzhou, China

Irene LU (IL), Assistant Curator, China National Silk Museum, Hangzhou, China

LU Jialiang (LJ), Assistant Curator, Zhejiang Sci-Tech University, Hangzhou, China

LUO Qun (LQ), Senior Researcher, China National Silk Museum, Hangzhou, China

Assadour MARKAROV (AM), Professor, China Academy of Art, Hangzhou, China

Linda MCINTOSH (LM), Independent Researcher, Luang Prabang, Laos

Rasuljon MIRZAAHMEDOV (RM), Weaver, Margilan, Uzbekistan

Rastjouy Ardakani MIRZAMOHAMMAD (RAM), Weaver, Ardakan, Iran

Sandra NIESSEN (SN), Independent Researcher, Oosterbeek, the Netherlands

Binafsha NODIR (BN), Art Historian, Tashkent, Uzbekistan

Magdalena ÖHRMAN (MÖ), Senior Lecturer in Classics, University of Wales Trinity Saint David, UK

Bouakham PHENGMIXAY (BP), Weaver, Ban Natoum, Laos

Elena PHIPPS (EP), Lecturer, Department of World Arts and Cultures, University of California, Los Angeles, USA

Safidy RAHARIVONY (SR), Weaver, Analamanga, Madagascar

Arline RAVAO (AR), Weaver, Ambalavao, Madagascar

Sandra SARDJONO (SS), President, Tracing Patterns Foundation, Berkeley, California, USA

Guy SCHERRER (GS), Engineer and Conservator/Restorer of Historic Machinery, Saint-Romain-Lachalm, France

SIM Yeonok (SY), Professor, Korea National University of Cultural Heritage, Buyeo, R. O. Korea

Yanet SOTO (YST), Weaver, Centro de Textiles Tradicionales, Cusco, Peru

Tatyana TRUDOLYUBOVA (TT), Culture Program Assistant, UNESCO Tashkent Office, Uzbekistan

Gillian VOGELSANG-EASTWOOD (GVE), Director, Textile Research Centre, Leiden, the Netherlands

I Putu Cobby WIRYADI (IPCW), Cultural Historian, Pegringsingan, Bali, Indonesia

YOSHIMOTO Shinobu (YS), Professor Emeritus, National Museum of Ethnology, Osaka, Japan

YU Youde (YY), Weaver, China National Silk Museum, Hangzhou, China

YUMA Taru (YT), Weaver, Taiwan, China

ZHAO Feng (ZF), Director, China National Silk Museum, Hangzhou, China

Conference and Workshop Programs

Conference Program

31 May 2018

Keynote Address: Origins, Transmission, and Future   *Eric Boudot*

Development of Looms and Weaving Technology on the Silk Road   *Zhao Feng*

Cradle of Diversity: Looms and Weaving Techniques of Southwest China   *Christopher Buckley*

Looms for Warp- and Weft-Twined Weave in Japan   *Yoshimoto Shinobu*

Celebrating the Lao Loom: Exploring Creativity and Innovation   *Carol Cassidy*

Backstrap Looms in the Indonesian Archipelago   *Sandra Sardjono*

The Andean Loom and the Four-Selvedged Cloth   *Elena Phipps*

Threads That Bind: A Perspective on *Jaala* and Other Indian Looms   *Hemang Agrawal*

Uzbek Ikat: Traditional Loom and Weaving Technology   *Binafsha Nodir*

Iranian *Zilus* and the *Zilu* Loom   *Gillian Vogelsang-Eastwood*

The Warp-Weighted Loom in Ancient Europe   *Eva Andersson Strand*

Madagascar: The Survival of a Weaving Tradition Through Funeral Rituals   *Andrée Etheve*

Creativity and Innovation: Looms and Weaving Technology in Africa   *Malika Kraamer*

Weaving Figured Textiles: Before the Jacquard Loom and After   *Guy Scherrer*

Workshop Programs

1 June 2018

Ancient Europe: The Warp-Weighted Loom
    Eva Andersson Strand and *Magdalena Öhrman*
Taiwan, China: Weaving Technology of Traditional Taiwanese Loom
    Weaver: *Yuma Taru*
    Translator: *Ho Zhaohua*
Cusco, Peru: Backstrap Loom
    Weavers: *Flora Callanaupa* and *Yanet Soto*
    Translator: *Elena Phipps*

R. O. Korea: Backstrap Treadle Loom
    Weaver: *Bang Yeonok*
    Translator: *Sim Yeonok*

Madagascan Loom
    Weavers: *Arline Ravao* and *Safidy Raharivony*
    Translator: *Andrée Etheve*

2 June 2018

Ghana: Frame Loom
    Malika Kraamer

Margilan, Atlas, Central Asia: Ikat Loom
    Weaver: *Rasuljon Mirzaahmedov*
    Translator: *Tatyana Trudolyubova*

Laos: Loom with Long Vertical Pattern Heddle System
    Weaver: *Bouakham Phengmixay*
    Translator: *Carol Cassidy*

China: Zhuang Loom from Guangxi Zhuang Autonomous Region
    Eric Boudot

West Sumatra, Indonesia: *Songket* Textiles from Minangkabau
    Weaver: *Bernhard Bart*
    Translator: *Erika Dubler*

3 June 2018

South Sumatra, Indonesia: *Songket* Loom from Palembang
    Weaver: *Zainal Arifin*
    Translator: *Sandra Sardjono*

China: Laoguanshan Pattern Loom
    Weaver: *Yu Youde*
    Translator: *Luo Qun*

Iran: *Zilu* Loom
    Weaver: *Rastjouy Ardakani Mirzamohammad*
    Translator: *Ata Jallayer*

Varanasi, India: *Jaala* Loom
    Weavers: *Naseem Ahmad* and *Tauseef Ahmad Ansari*
    Translator: *Hemang Agrawal*

France: Jacquard Loom
    Guy Scherrer

Contents

Introduction

1. Origins, Transmission, and Future — 5
   Eric Boudot

2. Classification of Looms — 13
   Zhao Feng

East Asia

3. Early Chinese Looms — 19
   Long Bo

4. Cloth Production in Ancient Tianluoshan — 23
   Judith Cameron

5. Chinese Imperial Workshop Looms — 29
   Long Bo

6. Minority Looms of Southwest China — 33
   Christopher Buckley

7. Japanese, Ainu, and Korean Looms — 45
   Yoshimoto Shinobu

8. Voice of the Weaver: Hangzhou — 62
   Yu Youde, Luo Qun

Southeast Asia

9. Mainland Southeast Asian Looms — 67
   Linda McIntosh

10. Insular Southeast Asian Looms — 81
    Sandra Sardjono

11. Voice of the Weaver: Palembang and Muang Kham — 94
    Zainal Arifin, Bouakham Phengmixay

South Asia

12. Indian Looms — 99
    Hemang Agrawal

13. Voice of the Weaver: Varanasi — 110
    Naseem Ahmad

Central and Southwest Asia

14. Central and West Asian Looms — 115
    Gillian Vogelsang-Eastwood

15. Traditional Weaving of Uzbekistan — 121
    Binafsha Nodir

16. Voice of the Weaver: Margilan and Ardakan — 130
    Rasuljon Mirzaahmedov, Rastjouy Ardakani Mirzamohammad

Africa

17. African Looms — 135
    Malika Kraamer

18. Madagascan Looms — 145
    Andrée Etheve

19. Voice of the Weaver: Analamanga, Ambalavao, and Tema — 148
    Safidy Raharivony, Arline Ravao, Bob Dennis

Europe

20. Ancient European Looms — 155
    Eva Andersson Strand

21. Voice of the Weaver: Two Ancient Tales — 160
    Magdalena Öhrman

The Americas

22. The Andean Loom and the Making of Four-Selvedged Cloth — 167
Elena Phipps

23. Voice of the Weaver: Cusco — 176
Flora Callanaupa, Yanet Soto

The Jacquard Loom and After

24. The Development of the Jacquard Loom — 181
Guy Scherrer

Catalog

East Asia: Archaeological Evidence from China — 191

East Asia: Present-Day Looms from China — 210

East Asia: The Korean Peninsula — 231

Mainland Southeast Asia — 232

Insular Southeast Asia — 240

South Asia — 251

Central and Southwest Asia — 256

Africa — 264

Europe — 275

The Americas — 279

The Jacquard Loom and After — 281

| Glossary | 285 |
| Bibliography | 292 |

Introduction

1. Origins, Transmission, and Future
   Eric Boudot

2. Classification of Looms
   Zhao Feng

1. Origins, Transmission, and Future

Eric Boudot

Weaving has been an integral part of life since at least the Neolithic Period. The ancient origins and the importance of weaving began to receive scholarly attention during the last decades of the 20th century. Since then, the number of publications on archaeological textiles and looms has been rising exponentially. Studies of ancient fabrics and weaving technology are now recognized as key for understanding the evolution of human culture over the past 10,000 years.

Weaving is one of the earliest technologies that necessitated mathematical thinking, associated with planning and warping complex textiles, and weaving prompted some of the earliest examples of mechanization. Many early weavings attest to technological expertise of a high order. Take for example the process to weave a 2,300-year-old polychrome warp-faced compound silk (figs. 1.1–1.3). The fragment—part of a coffin cover attributed to the Zuojiatang site in Hunan province, China—has 6,500 warps of 3 different colors in a width of only 51 cm. In other words, the weave has an extremely high warp count of more than 13 ends per mm. Imagine the precision it took in organizing the 6,500 silk threads in order to plan the intricate design: each warp thread had to be individually arranged following a certain order of colors and manipulated by one of the many thousands of heddle loops. In an age before calculators and computers, this process was truly remarkable. It required sophisticated thinking to group and arrange a large number of warps, control them with heddles, and to plan the interlace of warp and weft for the finished design.

The relationship between mathematics, technology and weaving has continued to develop until recent times. The precise control demanded by commercial silk weaving led to the development

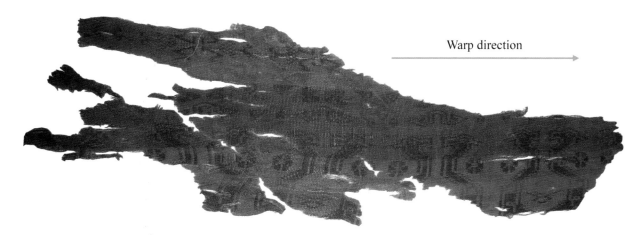

Fig. 1.1 Fragment from a Chu coffin cover, silk polychrome warp-faced compound plain weave, early 3rd century BCE, 24.3 cm x 6.15 cm. Private collection.

Facing page: *jin*-silk with geometrical patterns excavated from tomb 1 at Mashan, Jiangling, Hubei. Silk polychrome warp-faced compound plain weave, Warring States Period. Collection of Jingzhou Museum.

Fig. 1.2 Detail of fig. 1.1.

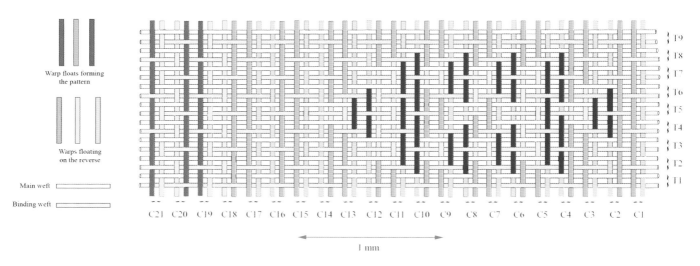

Fig. 1.3 Structure diagram of the detail in fig. 1.2.

of a punch-card control system as embodied in the Jacquard loom in France (section 24). A technology that had previously been a curiosity, used to control mechanical automata and other amusements, became the basis of an important industrial weaving technology. Today, programmed control systems are taken for granted, but all of these systems owe a debt to the engineers who worked on these early looms.

The high level of skill that weaving demanded also had social consequences. In the 18th and 19th centuries in France, silk weavers in Lyon were at the forefront of the fight for the worker's rights, and some historians relate their early social consciousness to the growing status of master weavers, based upon their understanding of the complex mechanisms of their looms. Napoleon recognized this situation and established the first tribunal dedicated to labor disputes in 1906 in Lyon.

1.1 The Origins and Conservatism of Weaving Traditions

In traditional settings, weaving is generally taught by mother to daughter through years of apprenticeship, and thus textile technique is, by nature, conservative. Such conservatism allows researchers to observe a wide variety of ancient textile production processes "in vivo." Field researches among the ethnic groups of Southwest China, for example, illustrate that the weaving of traditional patterned textiles has strict rules that must be followed faithfully from one generation to the next. This phenomenon can be observed when comparing the weaving of the Miao from Huishui with the Maonan from Libo in Guizhou province. The Miao people use one of the most rudimentary looms to make supplementary-weft patterned plain weave at a speed of about 1 cm a day. Meanwhile, less than 200 km away, the Maonan women use a more sophisticated loom to weave a fabric with a similar structure at a much faster rate of up to 50 cm daily. Even though these groups have probably been neighbors for a very long time, the Miao weavers have never considered adopting the Maonan technology. It is obvious that for the Miao weavers the end goal is not speed of production but rather something else, which is linked to the role of weaving in the Miao society and traditional and customary rules.

In South China, the preservation of weaving traditions is remarkably represented by at least 20 different types of looms. Simple body-tensioned looms are used by the Li people on Hainan island, for example, and it appears that simple looms with related tools were in use in the Hemudu Culture, located in the Yangtze Delta from at least 7,000 years ago, with the body-tensioned loom itself in use by the Liangzhu Culture in southern China around 5,000 years ago (cat. 3). In Southwest China, in particular, the mechanical differences shown in the loom varieties are exceptional: each represents an evolutionary step in the development of Chinese looms technology. The rare findings of active archaic looms in South China further allows scholars to confirm and fill in some major gaps in loom technological evolution.[1]

Another example of conservatism is the Egyptian horizontal loom of Bedouin nomads in the Libyan Desert. Such loom was illustrated on a ceramic bowl found in a tomb in Badari from the early 4th millennium BCE (section 14). Conservatism in Scandinavia can also be seen in the use of archaic warp-weighted looms up to the 20th century (section 20). More examples can be found in the body-tensioned looms of Southeast Asia (sections 9 and 10) and Peru (section 22), in India with the *jaala* loom (section 12, cat. 36), in Iran with the *zilu* loom (cat. 40), and in other looms described in the catalog part.

1.2 The Relationship Between Woven Structure and Loom Technology

As the supplementary-weft patterned weaves of the Miao Huishui and the Maonan Libo people illustrate, textile structures, considered in isolation, provide little information about the loom on which a textile is woven. As is the case with Miao weaving, a complex textile structure can be woven on a rudimentary loom. Therefore, studies of textile structure must be integrated with ancient looms and those from living traditions in order to determine the relationships between looms and textile structures. An example of a successful integration of these aspects is a recent study in connection with the discovery of models of ancient looms in Laoguanshan (cat. 7). For a long time, scholars have noticed that Chinese warp-faced compound plain weave textiles, often called *jin* in literature, have a very short repeat in the warp direction. After the recreation of the Laoguanshan looms by the China National Silk Museum, Hangzhou, as described in the catalog entry, the reason for short pattern repeat and the repeating anomalies became clear.[2] The Laoguanshan loom has a multiple patterning harness system, each harness representing one row of weft insertion. There is a limitation in the number of the patterning heddles that can be physically accommodated in the loom body and consequently a limit to the length of the pattern repeat.[3]

The study of living and historical weaving traditions in Peru, on the other hand, illustrates the point that there is no rigid relationship between the woven structure and the loom type. In the case of Peruvian looms, as with Li Hainan looms mentioned already, weavers use extremely basic looms, consisting, when disassembled, of little more than a bunch of sticks. All pattern insertion is done by finger manipulation of warp and weft. The weaver visualizes and plans the entire piece and its relationship to the finished garment, which may include four selvedges and the neckline without using any mechanical aids (section 22). Simple looms do not imply simple weaving, and in fact the weaver's task becomes more complex in many cases.

1.3 New Research Directions in the Study of Loom Evolution

In most traditional societies, the evolution of loom technology follows an internal, also called "vertical" or generation-to-generation transmission. At the same time, technology may be influenced by neighboring weaving traditions, leading to exchange of ideas between communities, which is called "horizontal transmission." The latter becomes particularly important in the case of professional workshop weaving, where commercial imperatives drive technological change and imitation. In the study of loom evolution, therefore, both vertical and horizontal transmission must be considered and carefully unpicked. With this in mind, the following research directions seem worthy of further exploration.

1.3.1 The Link Between Linguistics and Weaving Technology

Linguists have shown that language similarities and differences reflect the peoples' history of migration and assimilation (or non-assimilation). Studies in Southwest China have revealed close links between loom technology and the linguistic group, with complex patterning systems unique to the Tai language group, for example. Loom technology, because of its conservatism, also represents a route in its own right to understanding this history. Reconstructing the origins and dispersal of the many ethnic groups now living in Southwest China, a dispersal that has given rise to a complex patchwork of ethnicities and weaving techniques, is an interesting and challenging example. One exciting new tool that has recently been added to the textile scholar's arsenal of social, cultural and historical methods is phylogenetic analysis, by which genealogies of looms and textiles may be reconstructed.[4]

1.3.2 The Evolution of Drawlooms

Drawlooms, with their programmed patterning systems, are a vitally important part of contemporary commercial weaving. Though they now appear in a modern, computer-controlled guise (section 24), they owe their existence to a long evolution that includes developments in China, Central Asia, northern India and the Middle East, and later in Europe. This field is ripe for further study, particularly as drawlooms seem to represent mankind's earliest attempts to make complex mechanical control systems. In the Far East, critical contributions must have been made in Chinese workshops environment between the 1st century BCE Laoguanshan patterning heddles loom (cat. 7) and the drawlooms that are attested a thousand years later in Song Dynasty paintings. One promising research avenue is to study the emergence and impact of these technologies, which seem to have been associated with a major shift from warp-faced to weft-faced compound weaves and from plain weave to twill ground structures between the 3rd and the 8th centuries in China.[5] Linked to this topic is the study of the contributions to drawloom technology of Chinese and Central Asian traditional weaving cultures. To shed light on the subject scholars need a better historical understanding of the evolution of the Iranian *zilu* and the Indian *jaala* drawloom (which is said to have been imported by Sogdian weavers around the 14th century).

1.3.3 Professional Workshops in China and Their Role in Loom Technological Innovation

While domestic weaving traditions in China are conservative, the professional weaving workshops—producing patterned textiles for the aristocracy—brought about radical changes because they focused on fashion and productivity. Sometimes, new technologies arrived with foreign weavers who were captured and enslaved in China. Important technology exchanges and "cross-pollination" also seem to have occurred between early kingdoms in China, prior to unification around 2,000 years ago. In the period from the 1st millennium BCE up to the Song Dynasty (960–1279) the invention of sophisticated loom mechanisms in China gradually emerged within professional workshops. Examples of such inventions include the multi-heddles patterning system (Laoguanshan loom, cat. 7), the multi-heddle and multi-treadle system (*Dingqiao* loom cat. 13) and the drawloom (cat. 22).

1.3.4 The Relationship Between the Chinese Drawloom and the European Drawloom

Drawlooms appeared in Italy around the 15th century. The numerous features shared by the Chinese drawloom and the European drawloom imply a close connection. If so, what was the route of this technology transfer from China to Europe? Did the Arabo-Andalusian drawloom, which was still in use in Spain until the late 15th century, and the Indian drawloom play a role?

1.3.5 The Relationship Between Early Chinese and South American Body-Tensioned Looms

Some scholars have suggested that the loom with heddles in Peru, which appeared around 2000 BCE, could have been an imported technology.[6] The possibility that there was a connection across the Pacific is intriguing. For this, one needs to study ancient trade routes, both terrestrial and maritime.

1.4 Weaving Traditions Today and in the Future

Traditional looms around the world have been disappearing at an alarming rate since the beginning of the 20th century. Recently, however, there are increasing efforts to save traditional weaving from oblivion and to provide new roles for weavers and their products.

One factor is a resurgence of interest in identity and tradition. A story from the Shidong area of eastern Guizhou province inspires hope. At one point in the past few decades, young Miao women had almost completely stopped weaving one of

Fig. 1.4 A twenty-year-old Miao woman, back in her village from a distant factory job, learning the intricate patterned weaving technique required for traditional baby carriers.

the most important components of their wedding costume, the bridal apron. Weaving these aprons—decorated with delicate patterns and figures—requires years of apprenticeship starting at the age of twelve. In the 1990s, many young girls migrated to China's coastal areas in pursuit of factory jobs, and thus abandoned weaving. Mass-produced aprons of synthetic fibers, as well as jewelry made of cheap metal, locally known as *baitong*, began to replace authentic items. By the late 2000s however, resurgence in Miao pride in their identity and traditions began to change this situation. There is now a demand for high-quality handmade costumes and jewelry using precious materials, such as silk and pure silver. Within a short period, an entire group of villages in a mountainous region have developed a successful "cottage industry," which allows skilled weavers to earn more locally than unskilled worker in a distant factory (fig. 1.4).

Government-subsidized projects, which aim at the economic integration of villages with contemporary creative arts and crafts, are also contributing to the survival of traditional weaving. A new trend is the eco-museum that is meant to protect and preserve ethnic identity. The Dimen Dong cultural eco-museum, for example, was built in 2004 in Liping county, southeastern Guizhou province. The museum encompasses eight relatively isolated villages of the Dong (Kam) ethnic group. These villages are voluntarily absent from the tourist circuit but welcome researchers and designers from around the world who are interested in how local crafts can adapt to the modern world.

Another route to preserving traditional crafts is exemplified by the work of Carol Cassidy in Laos. She has successfully developed an international luxury market for Lao silk products, while retaining traditional methods. Her project has now been in operation for over thirty years and is therefore one of the most sustainable examples of adapting traditional weaving to an international market. In recent decades many similar projects have appeared around the world, notably in South America, in Africa and in Southeast Asia. Examples of collaborative work with international designers are textiles recently produced by Maonan and Tujia weavers in Southwest China (fig. 1.5).[7]

All the above projects highlight the need to pay attention to and support traditional weaving practices, in a world where "change" is the new "constant." Studies of the transmission of indigenous weaving, which is oral and demonstrative rather than text-based, emphasize its fragility, and the difficulty of recreating a tradition once it is lost. Weavers, scholars, collectors, hobbyists, designers, overseas visitors—and readers of this book—all have a role to play in the preservation and future success of our world weaving heritage.

Notes

1 Boudot and Buckley 2015.

2 The loom was reconstructed by Mr. Luo Qun, technical engineer at the China National Silk Museum, Hangzhou. See Voice of the Weaver section 8.

3 This is in contrast with the later weft-faced compound weaves (plain weave or twill), where the repetition of the motif in the weft direction is a result of the cords and leashes system, characteristic of drawloom.

4 Buckley and Boudot 2017.

5 Boudot (forthcoming).

6 Bird 1979.

7 Collaboration between Eric Boudot, Magali An Berthon (a French textile designer), and Maonan weavers in northern Guangxi Zhuang autonomous region to produce contemporary textiles, and collaboration with Tujia weavers in northern Hunan province to weave textiles decorating contemporary furniture created by the English designer Peter Harvey.

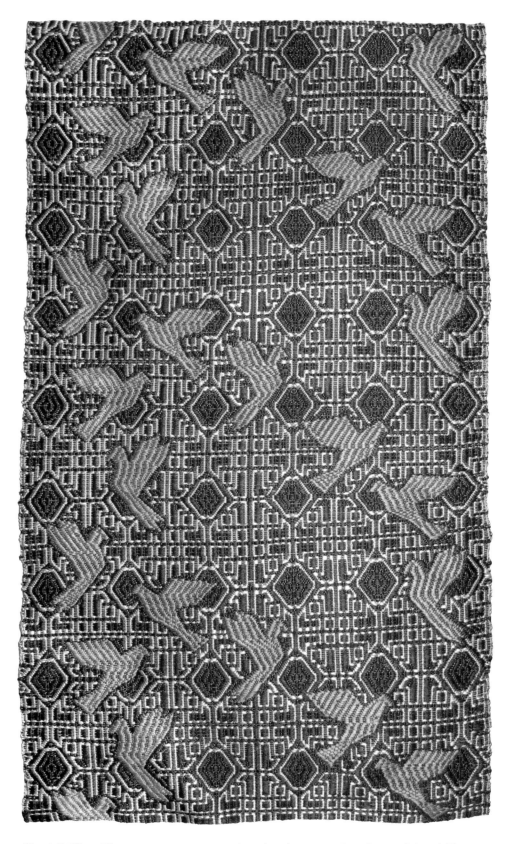

Fig. 1.5 "Envol," a contemporary creation, handwoven using the traditional Maonan technique. Handspun silk with natural dyes, 60 cm x 37 cm (Design: Magali An Berthon and Eric Boudot. Photo: Eric Boudot.).

2. Classification of Looms

Zhao Feng

Woven fabric consists of two sets of elements—warp and weft—that are usually interlaced at right angles.[1] The process of interlacing warp and weft is facilitated by a loom, a device for keeping the warp under tension and evenly spaced. The weft is inserted back and forth into alternating openings in the warp, called sheds. The weaving process can be broken down into five basic operations:

(1) Setting up a warp on the loom (warping).
(2) Creating warp openings (shedding).
(3) Inserting wefts into the sheds, usually using some kind of weft-carrier (spool or shuttle).
(4) Beating-in the weft, to make a firm "web."
(5) Releasing the completed fabric and finishing, for example by making fringes.

Of these five steps, creating warp openings is the most complicated, and much of the complexity of looms and the differences between them are concerned with this operation. The basic tool used on the majority of looms for opening sheds is a heddle. This consists of loops of cord, usually secured to a rod of some kind, which are attached to specific warps. When the heddle is lifted (or depressed, in some looms), openings are created in the warp, ready for weft insertion. The most basic loom designs have a rod (shed stick) that retains the "natural shed," and a single heddle that opens the opposite shed (counter shed), or a pair of heddles (called ground-weave heddles) that do the same job. Warps can also be selected directly by a weaver using a pointed implement, and this is the way that patterns are created on many simpler looms. Patterns can also be made using additional pattern heddles that record complex warp lifts.

The variations on these themes are endless, and for any given textile structure there are usually many different ways to make it using many different looms. A "simple" loom does not imply that the resulting textile is also simple: extremely complex interlaces may be created by a skilled weaver working with the simplest of equipment.

The basic warp-tensioning systems of looms vary widely: in some types the tensioning is achieved by using a frame or weights; in other types it is generated by the weaver's own body. Different solutions are characteristic of different regions. Some loom types may be restricted to a radius of a few kilometers, while others such as the Central Asian ground loom have spread over vast distances. Looms may be categorized based on whether or not they are equipped with features such as a frame, treadle, shaft, pattern device, shuttle, and electric power. For example, there are simple looms, frame looms, treadle looms, pit looms, pattern looms (of various kinds), shuttle-less looms, power looms and so on.

The simplest kinds of looms use three distinctly different systems for creating tension in the warp. In the warp-weighted loom, which was an important type in ancient Europe, the warp is hung from a top bar using weights to tension it. Weaving begins at the top of the loom and beating is done upwards, rolling the finished cloth onto the top beam. On two-bar looms, which can be vertical or horizontal, the warp beam and the cloth beam are fixed at the two ends of the warp. Versions of these looms are found worldwide. On backstrap looms, a type of body-tensioned loom, the cloth

Facing page: *jin* silk for an arm protector from the Han Dynasty (206 BCE–220 CE), bearing the woven inscription *wuxing chu dongfang li Zhongguo* (five stars rising in the east are auspicious for the Middle Kingdom) excavated from tomb 8 at Niya, Minfeng, Xinjiang. Warp-faced compound plain weave. Collection of the Xinjiang Institute of Cultural Relics and Archaeology (photo: Chen Long and Kang Xiaojing).

beam is attached to the weaver's waist while the warp beam may be braced using the feet or fixed behind stakes, or the warp may simply be tied to a stick in the ground, omitting the warp beam. In this type of loom the warp tension can be adjusted by the body movement of the weaver, facilitating the opening of shed and counter-shed. This type is most frequently encountered in East Asia, South Asia, Southeast Asia, and South America.

In treadle looms, heddles are attached to cords or bars operated by the feet; this system is more efficient than hand-operated heddles since it frees up the weaver's hands to work on weft insertion. The whole assembly of heddle-plus-treadle is usually called a shaft, and weavers talk about looms with two shafts, four shafts and so on. Treadles may be connected to the top of the shafts via a pulley system, or directly to the bottom of the shafts. In the first case, stepping on the treadles would lift the shafts, as in many East Asian looms. In the latter, the action would depress the shafts, as in a pit loom and many West African looms.

As mentioned, patterning can be created on any kind of loom through various shedding and patterning devices, or by a weaver patiently selecting warps by hand. Two major types of patterning system that are found worldwide are the multi-heddles system and the compound pattern leash system (drawloom system, fig. 2.1). Both of these systems can record complex designs so that they can be made easily and reproducibly. Many silk textiles found along the Silk Road display woven patterns produced by these two types of patterning system.

The evolution of complex loom patterning technology can be traced in broad terms as follows: first, multi-heddle looms appeared in the eastern part of the Silk Road (China), followed by the weft-faced-picking technique in Central or West Asia. Combined with multi-heddle system and weft-faced-picking technique, drawlooms developed in China during the Tang Dynasty (7th–10th centuries), thereafter somewhat different drawlooms appeared in Central Asia and North India. These looms were capable of recording extremely large and complex designs and were mainly used for luxurious silks.[2] Lastly, in the modern era the basic drawloom system was adapted and automated by the addition of punch-card control (Jacquard loom and related types) and most recently by full computer control of the weaving process. Such looms form part of commercial weaving technology that has now become a global industry.

Notes

1 Emery 1966.

2 See examples of the range of silks found along the Silk Road in Zhao 2016.

2. Classification of Looms

Fig. 2.1 Model of a Chinese drawloom.

East Asia

3. Early Chinese Looms
   Long Bo

4. Cloth Production in Ancient Tianluoshan
   Judith Cameron

5. Chinese Imperial Workshop Looms
   Long Bo

6. Minority Looms of Southwest China
   Christopher Buckley

7. Japanese, Ainu, and Korean Looms
   Yoshimoto Shinobu

8. Voice of the Weaver: Hangzhou
   Yu Youde, Luo Qun

3. Early Chinese Looms

Long Bo

3.1 Weaving Technology in China

China has a long and rich history related to weaving and the development of looms, with the first simple looms appearing more than 7,000 years ago. The Warring States and the Han Dynasty Periods (5th century BCE–3rd century CE) saw the evolution of treadle looms and multi-heddle looms, and this seems to have been a particularly important period for the development of looms with frames with advanced features that speed up the weaving process. The most sophisticated patterning loom, the drawloom, seems to have developed partly as a result of cultural exchanges along the Silk Road, which stimulated Chinese silk weavers to produce new types of polychrome textile with weft patterning in imitation of imported polychrome silks, and may also have included the introduction of new patterning technologies. The drawloom, the ultimate patterning loom of the pre-Jacquard era, seems to have achieved its final form by the Tang Dynasty (7th–10th centuries) and become the dominant system for weaving patterned silks from the Song Dynasty to the Qing Dynasty (10th–early 20th centuries). The UNESCO list of important cultural heritage in China includes items of both tangible and intangible heritage. Amongst these are excavated pattern loom models from the Han Dynasty as well as the sericulture and silk craftsmanship of China in general, which has continued to this day and includes the weaving traditions of both Han and other ethnicities.

3.2 Earliest Looms

The earliest looms in China appear to have been simple backstrap looms (cat. 1–3), where the warp is stretched between a fixed point and the body of the weaver. The foundation weave is usually plain weave. The two plain weave sheds (openings in the warp required to make the cloth) are called the natural shed and the counter shed. The first is opened by means of a shed rod over which one set of warps passes, and the second by a continuous string heddle, the loops of which can pull up individual yarns between the other set of warps. Weavers lean back and forth, using their body weight to adjust the tension of the warp. As they lean forward, the tension is relaxed and they can raise the heddle to open the counter shed. As they lean back, the warp becomes taut and returns to its natural shed, held open by the shed stick. As each opening is made, the weaver inserts a flat blade (sword) into the opening, turning the blade at right angles to enlarge the opening to allow the weft to pass through.

This loom has long since disappeared from the central regions of China, but simple looms of the same general type are still used by some minority groups in western Yunnan, as well as in many parts of Southeast Asia. The history of these looms in the region forms a vital part of the history of weaving in the entire region.

Facing page: plain weave silk fragment and silk threads excavated at the Qianshanyang site, Huzhou, Zhejiang, Neolithic Period. Collection of the Zhejiang Provincial Museum.

3.3 Treadle Looms

A treadle loom is a general term for a loom equipped with a pedal that is used to lift a shed opening device. The use of these devices changed the simple shedding system operated by hand to the treadle-lifting system operated by the weaver's feet. The new system frees up the hands from lifting warps and greatly improves the efficiency of weaving. Treadle looms first appeared in China during the Spring and Autumn and the Warring States Periods (770 BCE–221 BCE). The first visual depiction of such a loom (fig. 3.1), however, dates to the Eastern Han Dynasty (25–220). There are several types of treadle loom with different combinations of heddles and treadles.

Since very few complete looms are preserved from this period, we must rely on tomb models and murals, and some features of looms must be imagined and interpolated from the evidence available. What is certain is that by the Han Dynasty a variety of different frame looms were in use in China, including smaller looms used for domestic production (cat. 8 and 9) and larger looms, such as the Laoguanshan types (cat. 7), for workshop production of more complex textiles. The archaeological finds show that by the Warring States Period, possibly earlier, weaving in China had become differentiated into production for domestic use and production for official use, with different looms and weaving practices employed for each of these.

The first parts of the catalog entries (cat. 1–23) describe the development of Chinese looms from earliest times, from the remains of simple body-tensioned looms from archaeological excavations, to the impressive frame looms with multiple patterning heddles recovered from the tomb at Laoguanshan (fig. 3.2). Descendants of some of these looms have remained in use in remote countryside locations up to recent decades, and these looms can give valuable information for interpreting the archaeological evidence.

Fig. 3.1 Depiction of treadle looms on a stone relief excavated from Zengjiabao tomb in Chengdu, Han Dynasty. Collection of the Chengdu Museum.

Fig. 3.2 Model loom with multiple heddles recovered from the Laoguanshan tomb, Han Dynasty.

4. Cloth Production in Ancient Tianluoshan

Judith Cameron

During the Neolithic Period, several different loom types were developed in the northern and southern hemispheres that were based on different fibers and filaments. This essay describes the loom parts that were unearthed during the landmark excavations of the Tianluoshan site, one of the earliest archaeological textile sites surrounding Hangzhou Bay that were occupied when groups migrated from the middle Yangtze and settled in villages in the lower Yangtze Valley. The loom components from Tianluoshan call into question the dominant paradigm that the earliest type of Chinese loom was the foot-braced backstrap loom type that is depicted several millennia later on a famous bronze cowrie container (fig. 4.1) from Shizhaishan in Yunnan, dated to the Western Han Dynasty (206 BCE–9 CE). The evidence discussed here suggests that a simpler version of the backstrap loom may have been in use at the Tianluoshan site along with the foot-braced type. The large assemblage of cloth production tools currently under analysis also provides new insights into the technological and organizational complexity achieved by the occupants of this early Neolithic site around 7,000 years ago.

Fig. 4.1 Workshop scene depicting weavers using foot-braced backstrap looms on the tympanum of a bronze drum from a tomb at Shizhaishan (after Rawson 1983, plate 14).

Facing page: workshop scene depicting weavers using foot-braced backstrap looms on the tympanum of a bronze drum, from a tomb at Lijiashan, Jinning, Yunnan. Collection of the Lijiashan Bronze Museum.

4.1 The Site

The excavations of the deeply waterlogged Tianluoshan site (7000–6000 BP)[1] yielded more than 400 well preserved textile tools, which were found in close association with architectural structures estimated to house around 100 individuals (fig. 4.2).[2]

4.2 The Tools

Spinning, the first stage in a *chaîne opératoire* in cloth production, was indicated by precisely 300 spindle whorls used in fiber processing. Spindle whorls first appear when the site was occupied, increase in numbers in layer 6 and decrease in numbers in the final stage of occupation when water levels rose and the occupants presumably relocated (fig. 4.3).

Several different whorl types—discoid, conical, biconical, globular (fig. 4.4)—were identified along with truncated cones with narrowed waists providing space for thread to be wound around the tool, a prehistoric precursor to the modern cotton reel. The wide range of types suggests that different fiber-producing plants were spun at the site. Pollen and ethnographic studies suggest that *Urtica diocia* (nettle), *Pueraria thunbergiania* (bean creeper), *Boehmeria nivea* (ramie), *Cannabis sativa* (hemp), *Broussonetia sp.* (paper mulberry), and

Fig. 4.2 Excavations at the Tianluoshan site (photo: Sun Guoping).

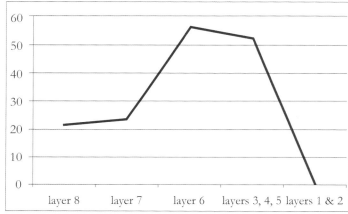

Fig. 4.3 Distribution of spinning tools at the Tianluoshan site.

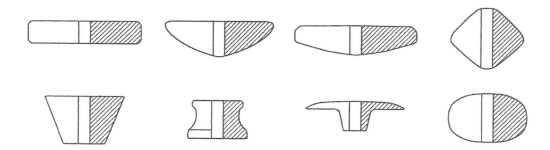

Fig. 4.4 Typology of spindle whorls from Tianluoshan.

Bambussa sp. (bamboo) were spun. Tree cotton (*Ceiba petanda*) and silk (*Bombyx mori*) may also have been spun.

Weaving—the next stage in the *chaîne*—was evidenced by wooden components of backstrap looms. These remains included small pegs (fig. 4.5) that appear to have been driven into the ground to hold warp threads under tension as weavers still do in West China and Japan (fig. 4.6).

The excavations also yielded wooden tubes, some of which were beautifully lacquered, that could have functioned as shed sticks used to open spaces for the spools holding weft threads to pass through. Spools (fig. 4.7) were also recovered from Tianluoshan. Archaeological correlates of the tubes were also found at the Luobowan site in Guangxi and are also amongst the components used for weaving unspun bark fibers into mats in Japan.[3] Significantly, a fragment of reed (*Phragmites australis* Cav.) matting dated to 6775–6645 cal. yr. BP was also found at the site, which is the earliest directly dated woven artifact in China.[4] The end beams on looms used by the Montagnards in South Vietnam and other minorities are made of narrower tubes, often bamboo, as seen on the loom used by the Li from Hainan island. The site also yielded thin sticks like those functioning as shed sticks on contemporary Hainan looms.

The most unexpected components in the Tianluoshan assemblage were the eight warp spacers: ingenious devices with grooves designed to keep warp threads in place (fig. 4.8). They would have been necessary on the looms anchored to the ground with pegs as this type does not have an end beam. Warp spacers such as these are still used in many traditional looms, including looms in Borneo.

Fig. 4.5 Warp pegs, wood, approximately 20 cm long, Tianluoshan (photo: Judith Cameron).

Fig. 4.6 Left: Yunnan loom. Right: Ainu looms showing the peg anchoring the warp to the ground (drawing after Roth 1934, figs. 5 and 7).

Fig. 4.7 Spool, Tianluoshan.

Fig. 4.8 Warp spacer, Tianluoshan, 26 cm x 1.6 cm (photo: Judith Cameron).

Fig. 4.9 Left: wooden beater, Tianluoshan. Right: Hainan island loom (after Stübel 1937, plate 42).

Two different types of beaters were used during the weaving process to straighten weft threads and to beat them in. The large wooden sword-beaters were most probably used for beating-in bast fibers (fig. 4.9) whereas the large numbers of smaller highly standardized, intricately carved and polished bone beaters made from deer antler and water buffalo horn, were probably used for matting fibers. Most of the latter were broken when found. Thread was probably threaded through the holes drilled in them and hung from the waist.

The final stage in the Tianluoshan *chaîne* is represented by bone needles (fig. 4.10). The longer needles with large eyes resemble Palaeolithic examples with large eyes first occurring in China at caves like Zhoukoudian and Xianrendong, whilst smaller needles with eyes placed near their ends and measuring less than 1 mm in diameter were probably used to sew finer cloth and gauze, or for adding embroidery, which is one of the hallmarks of the textiles made by Chinese minorities today. During the dynastic period, embroidery reached

Fig. 4.10 Bone needles of various sizes, Tianluoshan (photo: Judith Cameron).

such a high standard that it was prescribed for the nobility and the earliest examples occured in royal graves. The Prince of Chu, for example, so loved his horse that it was dressed in embroidered cloth. Sima Qian's *Shiji* (*Records of the Grand Historian*) describes the use of embroidered cloth in diplomacy. The Prince of Zhao sent 100 horse-drawn chariots, 200 jade tablets and 1,000 reels of embroidered silk to fill the campaign trunks of Zhao and establish an alliance amongst the six states against Qin.

4.3 Conclusions

The level of technological complexity represented in Tianluoshan's cloth production tools far exceeds what was previously known from early Neolithic sites. The warp spacer, for example, is an important innovation in loom technology, designed to increase both operational efficiency and output, suggesting that lower Yangtze groups had moved beyond domestic production, to community specialization, a production mode characterized by complexity, consisting of household-based specialists concentrated in one settlement producing for regional consumption.[5] This suggests that the groups in the lower Yangtze were probably involved in trade and exchange during the prehistoric period, long before the Silk Road of the historical period.[6]

Notes

1 Sun and Huang 2004; Zhang et al. 2016.

2 The author was invited by the excavators to analyze the spinning tools from Tianluoshan, and subsequently collaborated with Long Bo (section 3) to investigate loom components before they were placed on display during the exhibition *A World of Looms* at the China National Silk Museum in 2018.

3 See Roth 1934, fig. 14.

4 Zhang et al. 2016.

5 Costin 1991, 8.

6 Cameron 2004; Cameron 2017.

5. Chinese Imperial Workshop Looms

Long Bo

A drawloom is a loom designed for making complex patterned textiles. It stores a large number of warp lifts, corresponding to pattern weft insertions, in a single set of pattern leashes, thus preserving designs for repeated use. Most drawlooms have two sets of heddles: ground-weave heddles, located near to the weaver and operated by the weaver via treadles, and a separate pattern heddle, usually worked by a drawperson. The weaver uses the ground-weave heddles to make the basic textile structure (plain weave, twill, satin, etc.), while the drawperson lifts the warps for pattern weft insertions. The two must coordinate their movements closely.

There are several versions of the Chinese drawloom; each has slightly differing arrangements of patterning leashes. All the versions have a distinctive "tower" in which the drawperson sits, pulling the pattern leashes to raise the warps.

Designs are recorded as a set of cords embedded in the pattern leashes. The drawperson uses each cord in order, pulling it forwards to raise the leashes, then in the case of the Greater Drawloom removing and re-inserting the cord for future use. In some versions of the loom for making polychrome silks an extra set of leashes is used to indicate the color of weft that the weaver must insert. In this way a complete design is encoded, making the drawloom the first device that records patterns in "digital" form. The drawloom design overcomes the space limitation encountered by looms that use multiple heddles to record designs. The number of pattern cords that can be saved by the larger versions of the drawloom is essentially limitless. The pattern cords can also be saved and transferred from one loom to another with less effort than is needed to change the patterning in multi-heddle looms.

Shown here (fig. 5.1) is the earliest extant illustration of a drawloom, taken from a handscroll from the Song Dynasty (12th century). The handscroll, called *Canzhi Tu* (*Pictures of Silkworms and Weaving*), shows a loom with a pattern tower rising above the middle of the loom. The pattern system suspended on it consists of: "foot threads," which are the harnesses that connect directly with the warp, the threads of the pattern draft; and "ear threads"—that is, the horizontally arranged cords of the pattern draft that record the design. The foot threads pass through a grid of wooden rods that keeps bunches of threads in their correct positions, to weights hanging below the tower that tension the threads. The drawperson sat on top of the tower and used both hands to pull the

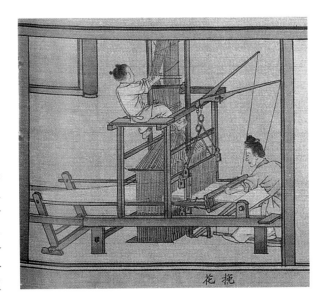

Fig. 5.1 Drawloom from the *Canzhi Tu* handscroll, annotated by Empress Wu, Southern Song Dynasty, 12th century, 27.5 cm x 513 cm. Collection of the Heilongjiang Provincial Museum.
Facing page: polychrome silk with pattern of lion inside flower medallion, in the style of the Duke of Lingyang. The weave is a specific type of weft-faced compound twill, often called the "Liao Dynasty" samit, Tang Dynasty (7th–10th centuries).

appropriate pattern cords. The weaver operated the treadles, working the large swinging reed-beater with one hand while passing the shuttle with the other. The beater is set in a heavy frame to increase the beating force. A double warp beam (fig. 5.2) at the back was used to reduce friction in the warp and create a better shed when the weave was particularly dense. This arrangement also allowed the lengths of the ground warp and the pattern warp to be regulated separately.

The final evolution of drawloom technology occurred during the late Tang Dynasty with the emergence of the Greater Drawloom, represented by the lampas drawloom and the *Yun* brocade drawloom (cat. 22 and 23). These looms are characterized by the large size of the pattern repeats, which can include as many as 100,000 pattern cords. A set of circular pattern cords, connected to the pattern harness, serves as an extension of the pattern harness and holds the pattern cords. As with other drawloom designs, each pattern cord is transferred, after use, from one side of the pattern harness to the other side and stored until the next repeat.

The most complicated and specialized task associated with the drawloom is making new sets of pattern cords (fig. 5.3) for new designs. Regarding this process, Song Yingxing provided the classic description in *Tiangong Kaiwu* (《天工开物》) or *Exploitation of the Works of Nature*:

"The artisan who makes new figure designs for weaving must be clever. An artist first paints the pattern and color of a fabric design on a piece of paper, then the artisan examines the design, then figures out how to render it in silk. The new design is first hung on the 'figure tower' of the drawloom, even if the artisan doesn't know exactly how the design will appear in the end. The appearance of the design in damask and thin silk depends upon the weft. In the weaving of damasks and pongees, the warp is lifted at every insertion of the shuttle, but in the case of gauzes it is lifted only at every other insertion. The weaving skills of the Heavenly Weaving Goddess[1] have truly been mastered by skillful artisans here in the country."

There are usually three processes involved in making new sets of pattern cords: picking, copying, and piecing the pattern. "Picking" the pattern

Fig. 5.2 Drawloom with a double warp beam from a Song copy of *Gengzhi Tu* (*Pictures of Tilling and Weaving*), Southern Song Dynasty, 12th century. Collection of the National Museum of China.

5. Chinese Imperial Workshop Looms

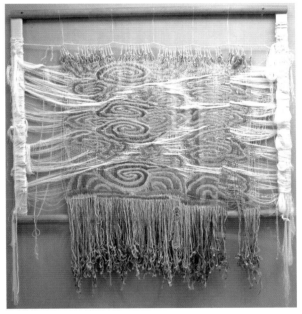

Fig. 5.3 Above: cloud pattern design drawn on graph paper. Each color corresponds to a pattern weft.
Below left: construction of a set of pattern cords. The wefts for the design have been inserted in their actual colors in order to assist the pattern maker.
Below right: the finished fabric in brocaded satin with polychrome cloud pattern.

means to figure out the basic interlace of warps and wefts. In case of a larger pattern, it is possible to save time and expense by using the method of copying or piecing patterns, re-using certain repeated elements.

The pattern-cords system was used for a long period in history, but it was gradually replaced by punched cards after the invention of the Jacquard machine. The Jacquard machine uses a binary code: if there is a hole, the warp moves up; if there is none, it stays in place. Different sets of holes of the punched cards encode different patterns. In essence however, the Jacquard loom (section 24) works on the same principle (recording warp lifts) as the older drawloom. Later, punched cards were also used in early telegraphy and in mechanical computing. Hollerith punched cards, resembling Jacquard cards, were used for storing computer programs until the 1970s and played an important role in the development of modern computing.

Notes

1 The Weaving Goddess (织女星) is identified with the star Vega in Chinese mythology.

6. Minority Looms of Southwest China

Christopher Buckley

China is an ethnically diverse country, with a particular concentration of minority groups in the Southwest, comprising Guizhou, Guangxi, Yunnan and parts of the neighboring provinces. This region includes speakers of Tai, Miao-Yao, Austroasiatic and Tibeto-Burman languages, as well as Han.

Until recent times weaving played an integral role in daily life for most of these peoples, with locally-made clothing signaling group identity, marital and social status. Fine weavings were required for rites of passage, including wedding trousseaus and funeral rituals (fig. 6.1). As in other parts of Asia, most weavers were female and weaving prowess was linked to the social status and marriageability of women, as well as the pleasure and pride in wearing beautiful, handmade clothing.

Given the variety of languages spoken, it is not surprising that there is a variety of looms used by China's minority peoples. What is surprising, however, is the depth and extent of this variety. In a relatively small area in the Southwest nearly all of the major types of loom that are used across the Asian region can be found. The range extends from the simplest kinds of body-tensioned looms, composed of little more than a few sticks, to complex programmed pattern looms with rigid frames and hundreds of components. In spite of this variety, the looms of this region share some characteristic features. In nearly all looms the warp

Right: fig. 6.1 detail of a bedcover made by a Buyi weaver from Guangxi, using a loom with a programmed patterning device. Silk and cotton supplementary-weft patterning on a cotton foundation. Private collection.

Facing page: detail of a bedcover woven by a Buyi weaver from Guangxi Zhuang autonomous region in China, using a loom equipped with a programmed patterning device. Private collection.

is oriented horizontally. Body-tensioning (the warp is tensioned by the weaver's own body) is a common feature. Many weavers use a long "sword," an all-purpose tool for making openings in the warp and beating-in wefts. These features and others like them offer clues to the interrelationships of the looms.[1]

Under the heading of "looms with frames" there is also a variety of types. These include several different systems for making openings in the warp and holding them open by means of treadles to raise heddles (shafts). These are devices that speed up weaving and free up the weaver's hands to work on weft insertion. Three distinctly different types of treadle (discussed below) are found in this region, each of which seems to have a long history.

The languages and looms are linked to the origins of the peoples. In the distribution of looms we can trace migrations and social contacts over a period of at least three millennia. The discovery and gradual development of rice farming in Central China and South China made new agricultural lifeways possible, in which weaving seems to have played an integral role. Austronesian speakers, who are still present in Taiwan today, originated in the southern coastal region of China, eventually spreading to many of the Pacific islands, carrying their body-tensioned ground looms and weaving traditions with them. Similar looms are still used by distantly related peoples in the upland regions in Yunnan province and on Hainan island. Tai peoples seem to have originated in Guangxi Zhuang autonomous region, which remains the center for the Tai language and textile diversity. They have distinctive and complex frame looms and patterning traditions, which they took with them on their migrations into Southeast Asia. Miao-Yao (Hmong-Mien) peoples similarly took their body-tensioned frame looms equipped with a single foot-treadle on their migrations into what is now Vietnam and northern Thailand. Their textiles and techniques in these regions have strong links to those of their cousins in South China.

6.1 Frameless Body-Tensioned Looms

The simplest looms found in Asia are body-tensioned types where the warp is wound in a circular fashion between two beams. A cloth beam is attached to the weaver's waist by a strap. A warp beam is either braced with the weaver's feet, or lodged behind two stakes fixed in the ground, or fixed in some other way such as to the side of a house (fig. 6.2). This loom is also equipped with a heddle that is raised by hand to open the counter-shed, and a shed stick that retains the natural warp

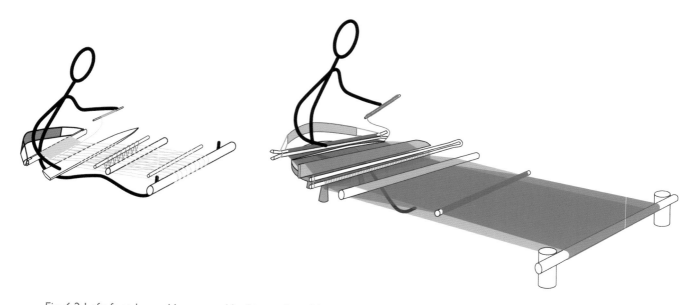

Fig. 6.2 Left: foot-braced loom used by Li people in Hainan.
Right: externally braced loom used by Wa people in Yunnan (drawings: Christopher Buckley).

Fig. 6.3 Li weaver in Donghezhen in Hainan, weaving cloth decorated with warp ikat, using a foot-braced body-tensioned loom (photo: Christopher Buckley).

opening. A hardwood sword is used for opening warps and beating-in wefts. Some weavers also use a coil rod to keep the warps in order, and other aids such as extra heddles for patterning are sometimes added. The shuttle for inserting the weft is often just a stick with the yarn wound around it.

Looms with the warp beam braced by the feet are used by Li weavers in Hainan (fig. 6.3), and by Austronesian speaking weavers in Taiwan (cat. 14).

This foot-braced loom is probably the most ancient loom in Asia:[2] there is evidence that it was once widespread in southern China, shown by the remains of a jade loom from the late Neolithic Period (cat. 3) and by bronze loom parts from Dian Culture burials dating from around 2,000 years ago found in Yunnan (cat. 5). Evidence has also been found of its use in southern Japan around the same period. Aside from the Li and

Fig. 6.4 Skirt woven by a Meifu Li weaver from Hainan, on a foot-braced body-tensioned loom (see fig. 6.3). The skirt is made from four strips of cloth woven separately, joined together, 60 cm x 93 cm. The warp runs horizontally in this photograph. Decorative weaving techniques include warp ikat, warp patterning and supplementary weft in silk on a cotton ground. This kind of skirt was worn by an unmarried woman at weddings and festivals. Private collection.

Taiwan weavers, this loom is still used today by a few Austroasiatic speakers in Laos and southern Vietnam (detailed in section 9.1). The wide geographic scatter of this loom, and the variety of peoples and languages using it suggest that it is a particularly ancient type that pre-dates the dispersal of farming peoples in Asia.

A slightly more sophisticated version of this loom with the warp beam fixed externally is used by some Tibeto-Burman and Austroasiatic language speakers, mainly living in the mountainous border region between Yunnan and Burma and on the Qinghai-Tibet Plateau, such as the loom used by the Wa people in fig. 6.2. For Tibetan weavers this loom is one of several in their repertoire, which includes upright looms used for making blankets and rugs, and ground looms fixed to the ground at four points with stakes or heavy stones (a Central Asian type, cat. 38 and 39), used for making tent bands. The Tibetan body-tensioned loom is mainly used by nomadic groups for making lightweight textiles. For some Austroasiatic speakers, such as the Wa, this is their only loom. Similar looms are used by related groups across Myanmar and the Himalayan foothills into Assam in India. This loom is also essentially the same as the one used by weavers speaking Austronesian languages across the islands of the Philippines and Indonesia, where it is particularly associated with textiles decorated with ikat, a technique in which the Li of Hainan excel (figs. 6.3 and 6.4). Geographically speaking, it is one of the most widely dispersed looms, being found from remote Pacific islands in the east (section 10) to Madagascar in the west (section 18). In Madagascar it was brought by Austronesian speakers around a millennium ago and is now a rare type used only at the southern tip of the island. As with the foot-braced version, the wide distribution of this loom implies a long history.

With the warp and weft removed, these simple looms are little more than bundles of sticks. They can be assembled quickly and with few resources and they are compact and portable: at the end of the day the weaver can simply roll the loom up. These advantages seem to have ensured their survival, particularly in marginal and resource-poor regions, even as they were displaced by more sophisticated looms in prosperous lowland areas.

Fig. 6.5 Miao weaver from the Huishui area of Guizhou province, using a half-frame loom to weave a cloth for funerary use, decorated with supplementary-weft patterning (photo: Eric Boudot).

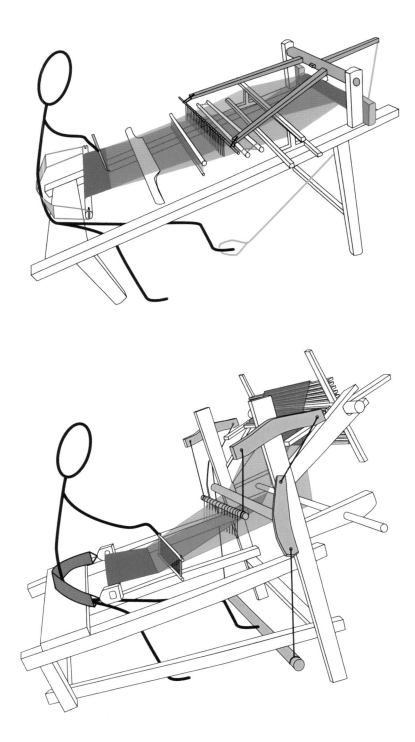

Fig. 6.6 Above: half-frame loom with a long rocker, used by Miao weavers from the Geyi area of Guizhou province. The I-shaped warp beam is shown in brown, and the mechanism for raising the heddle, consisting of a cord around the weaver's foot linked to a long rocker, is shown in green.

Below: half-frame loom with a short rocker, used by Zhuang people living near the border between Guangxi of China and Vietnam. The mechanism for raising the heddle is shown in orange. Versions of this loom are used by Tujia weavers in Hunan and by Han Chinese weavers in rural areas (drawings: Christopher Buckley).

6.2 Body-Tensioned Looms with Half-Frames

At some point in the history of weaving, the simple body-tensioned loom just described was placed in a rigid frame, raised on four legs and with a plank across one end for a seat. The initial result of this innovation might have looked similar to the loom that is still used by Miao weavers in the Huishui area of Guizhou (fig. 6.5). The Huishui loom also incorporates two more advances: a flat warp that is wound onto the warp beam at the start of the weaving, and a reed. The flat warp allows a longer length of warp to be mounted on the loom, 10 m or more being the norm on most looms with frames, the finished cloth being wound onto the cloth beam as the weaving progresses. The reed keeps the warps evenly spaced and allows better control of the warp spacing, which is helpful for a weaver who wants to make her weft visible, for example if she is making weft patterning.

Despite these advances, this remains a body-tensioned loom, quite different in design and conception to full frame looms with fixed cloth beams that are familiar to professional weavers and hobbyists in the West. Its origins are still apparent in the arrangement of the warp beam, which is not permanently fixed in the frame but rests behind two supports. When the warp beam and cloth beam are removed and rolled up, as weavers often do at the end of the day, the resulting bundle looks little different from the older, frameless looms.

The reasons why weavers originally chose to add a rigid frame with a raised seat are unknown, but once this advance had been made it opened up new (and initially unforeseen) possibilities. The half-frame allowed weavers to suspend the heddle (or heddles) from a fixed point on the frame, and to connect it to treadles. This allowed shed and counter-shed to be opened by treadle, speeding up the weaving operation. There exist two different body-tensioned looms with treadles that seem to have developed independently and that have different ethnic associations: a loom with the heddle attached to a long rocker, and a loom with a short rocker and a warp depression bar.

6.3 Half-Frame Loom with a Long Rocker

This loom is particularly associated with Tai and Miao speakers in the southwest. Several versions of this important loom, used by Dong, Tai and Zhuang weavers are shown in this catalog (cat. 15–17). The cloth beam is attached to the weaver's waist by a backstrap, while the I-shaped warp beam is lodged behind two uprights in the frame. In the simplest form of this loom (fig. 6.6, above) a single heddle is attached to a long U-shaped or Y-shaped rocker at eye level, which connects to a cord around the back of the loom that is attached to the weaver's foot. Pulling the cord raises the heddle and opens the counter-shed. Releasing the tension restores the natural shed, which is retained by a shed stick (a wide bamboo tube or a frame) that sits permanently in the warp.

The three versions of this loom in this catalog, which are all from Tai speaking weavers, also include a pattern heddle attached to a second rocker. The pattern heddle records the warp lifts for supplementary weft designs. Many Miao weavers also use this loom, though their patterns are selected by hand or by reference to a sampler, rather than by using a pattern heddle (fig. 6.7).

In China this loom is found exclusively in the southwest. It was carried into Southeast Asia by migrating Tai and Miao peoples and is now found across northern Vietnam, Thailand and Myanmar, as far west as Assam. To the east it is also found on the Korean Peninsula (section 7.42, cat. 24) and in Japan, where it is an important domestic loom known as a *jibata* (sections 7.26 and 7.27). As with the simpler body-tensioned looms, the wide distribution of this loom is evidence of a long history.

6.4 Half-Frame Loom with a Short Rocker

This loom uses a complex and ingenious system for opening the warp. A single heddle is attached to the loom frame by a short rocker, which is also attached to a transverse treadle and a warp depression bar (fig. 6.6, below). When the weaver

Fig. 6.7 Above: baby carrier made by Miao weavers from the Geyi area, with two panels of supplementary weft, finished with embroidery and appliqué.

Below: detail, showing supplementary weft designs in silk on indigo-dyed cotton ground.

Fig. 6.8 Loom for weaving ramie cloth, early 19th century. In this variant the cloth beam is fixed to the loom frame, necessitating a second treadle to restore the natural shed. Album leaf from an unpublished manuscript "The Story of Ramie from Seed to Finished Garment," 1820.

presses the treadle, the heddle is raised, pulling the lower warp layer through the upper and opening a counter-shed. At the same time, the warp depression bar holds the upper warp layer in place, which allows the weaver to make a wide opening in the warp. The natural shed is held open by bars that are permanently fixed to the loom frame. As with other body-tensioned looms, the weaver leans back to tension the warp when the natural shed is restored.

This loom is particularly associated with weavers speaking Sinitic languages (Chinese and closely related languages), including Tujia weavers in Hunan province, and Yao weavers in Jianghua county in Hunan, who speak a Han dialect. It is also used by a few Zhuang weavers in the border region between Guangxi of China and Vietnam. Aside from these minority peoples, this loom was formerly used by rural Chinese people all across central China (cat. 10), who used it until recently for making plain or striped cotton cloth for domestic use. It is an efficient loom for making plain cloth, which was surely a point in its favor for rural residents, who were required to produce rolls of cloth as a form of taxation prior to the middle of the 16th century. Unlike the long-rocker version, this loom does not seem to have spread significantly beyond the borders of China.

There are a number of variations on this basic design. One common variant that is found across China has a reed that is suspended overhead on two long, flexible strips of wood (fig. 6.8). Versions of this loom also exist in central China that fix the cloth beam into the frame. In this case, a second treadle is usually added to assist with restoring the natural shed, since the weaver can no longer use body tensioning for this purpose.

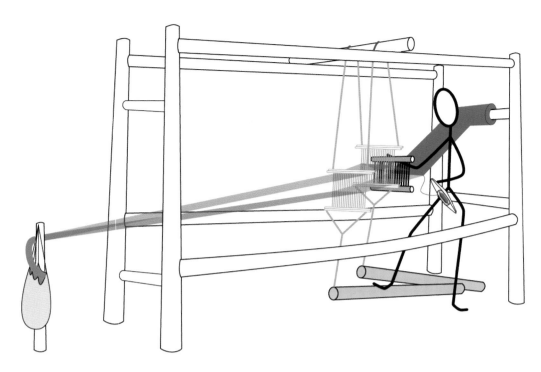

Fig. 6.9 Loom used by Hani people, Yunnan province (drawing: Christopher Buckley).

6.5 Full-Frame Looms with Fixed Cloth Beams

Most modern frame looms employ a pair of heddles, each of which can pull the warp in two directions (clasped or bidirectional heddles), linked to a pair of treadles. This allows the weaver to open shed and counter-shed rapidly, without needing body-tension to restore the natural shed. This allows the cloth beam to be fixed to the loom frame as well as the warp beam.

Looms based on this globally-important invention are found all across China (cat. 11 and 18), and have been slowly replacing older loom types. In many areas in the southwest this type of loom is used for making plain cloth, while the older body-tensioned looms are retained for making particular cloth with ceremonial importance, such as decorated bed covers for a wedding trousseau.

The simplest version of this loom is that used by some Yi, Lisu, Hani and other peoples living in upland areas of Yunnan of China and parts of northern Vietnam. The weaver stands upright while weaving, moving a small pair of treadles down the length of the warp while weaving (fig. 6.9, section 9.3, fig. 9.6).

In the most common version of the loom the weaver sits in front of her weaving, with the heddles and treadles suspended in front of her. This loom is essentially identical to the one that arrived in Europe from the 10th century onwards, displacing the older warp-weighted loom and two-bar frame looms that were previously the mainstays of European weaving. It is found in all parts of China. Looms employing the two-treadle system are also found in India (section 12.2.5) and across Central Asia (sections 14.3 and 15), Europe and North Africa (section 17.3). The loom eventually arrived in the Americas at the time of the Spanish conquests. Its origins are unknown, but the presence of some rare variants, such as a version used by Buyi weavers in Libo county in Guizhou that combine bidirectional heddles and treadles with body-tensioning, hint that it has had a long history in East Asia, and that its origins may even lie in this region.

6. Minority Looms of Southwest China

6.6 Patterning Systems

As mentioned, many Tai weavers use ingenious pattern heddles on their looms. These devices consist of a set of long leashes that control individual warps. Sticks or cords are placed in these leashes, some leashes passing above the stick (active leashes) and some below (passive leashes). The sticks are used in strict order: the weaver pulls a stick towards herself, which raises the leashes that lie above it and their corresponding warps. She inserts a colored pattern weft in the warp, and at the same time she removes the stick and replaces it in the mirror-image opening that appears behind the set of leashes that she has pulled up. She then repeats this process with the next stick. In this way the pattern is transferred to the woven cloth and can be used repeatedly.

These remarkable systems record a weaving design permanently. A design can be mounted on the loom, then reused, or taken off and a different design mounted. Most patterning systems consist of dozens of sticks, but larger pattern systems can contain hundreds. In Tai looms in southern Yunnan of China, Vietnam and Thailand, cords are used instead of sticks, and the patterns may run to hundreds or thousands of warp lifts. Patterning systems of this type can be seen on some Dong, Tai, Zhuang and Mulao looms (fig. 6.10), and on Lao looms (cat. 15–17, 26). The basic concept, recording warp lifts in a single set of leashes, is the same as that used to record patterns in silk drawlooms used in imperial workshops (cat. 20–23).

Fig. 6.10 Detail of a wedding bedcover made by a Mulao weaver, using a patterning system includes a permanent record of the warp lifts needed to make the design repeats. Silk supplementary weft on indigo cotton ground. Private collection.

Notes

1 Boudot and Buckley 2015; Buckley and Boudot 2017.

2 For a dissenting view see Cameron, this volume, section 4.

7. Japanese, Ainu, and Korean Looms

Yoshimoto Shinobu

Woven fabrics have been produced in Japan and on the Korean Peninsula. Some of their looms have been used to weave fabrics that are not regarded as "woven fabrics" by some authors. However, in this essay I define "woven fabric" as follows: woven fabric is constructed by combining weft threads with warp threads to which tension is applied. This is a broad definition, which includes textiles classified as made by the braiding technique by other authors. And the "Japanese looms" in this essay do not include "Ainu looms."

Among the looms described here are a body-tensioned loom which does not use any tools, a hand-tensioned loom, a foot-tensioned loom, a back-tensioned loom, a core-tensioned loom, a board-tensioned loom, a rod-tensioned loom, a hanging-weight tensioned loom and a frame-tensioned loom that does not have a shedding device.

7.1 Early Woven Fabrics in Japan

The oldest woven fabrics in Japan are fragments of warp-twined weave excavated from archaeological sites belonging to the early Jomon Period (5000 BCE–3500 BCE). The oldest fragments of patterned fabrics, made with a warp-twined weave, were excavated from sites dating from the late Jomon Period (4000 BCE–3000 BCE) in Hokkaido. In addition, some parts of looms such as warp beam, cloth beam, spool, and weft beater have been excavated from sites of the Yayoi Period (10th century BCE–3rd century CE).

The oldest known fragment of warp-twined weave in Japan was excavated from Torihama Shell mound, Wakasa, Fukui prefecture (fig. 7.1). The material of warp and weft is Boehmeria silvestrii (Pampanini) W. T. Wang, a type of nettle fiber.

One of the oldest fragments of patterned fabrics, with a warp-twined weave structure in Japan, was excavated from Kashiwagigawa 4 Ruin at Eniwa, Hokkaido (fig. 7.2). The material of warp and weft is a plant fiber, but the type of plant has not been confirmed.

Fig. 7.1 Fragment of warp-twined weave. Collection of the Fukui Prefectural Wakasa History Museum.

Facing page: modern copy of a Tang Dynasty silk with hunters and lions in weft-faced compound twill (samit), Japan, 1924. Collection of the Tokyo National Museum.

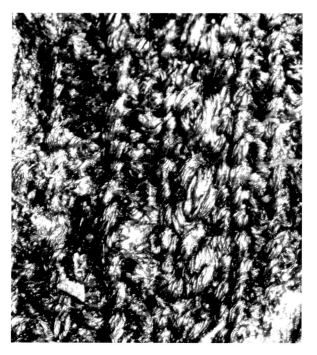

Fig. 7.2 Fragment of patterned warp-twined weave. Collection of the Eniwa City Local Museum, Hokkaido.

7.2 Japanese Looms

7.2.1 Body-Tensioned Loom

Sandal weaving (fig. 7.3) is done on a simple body-tensioned loom with the weaver's own body. No tools are used, and sandals are woven with only the feet and hands. The weaving technique is weft-rib weave, and the material for both warp and weft is rice straw. This picture of sandal weaving was taken at Nakanojo, Gunma prefecture in 2011. The sandal itself in the process of weaving is now in the collection of the National Museum of Ethnology, Osaka.

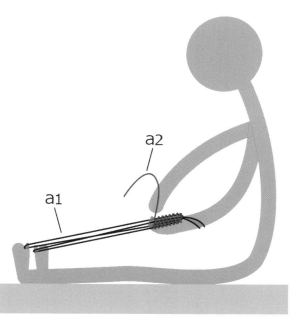

Fig. 7.3 Sandal weaving. National Museum of Ethnology, Osaka, H0270233; diagram: a1—warp, a2—weft.

7.2.2 Hand-Tensioned Loom

This picture shows another sandal weaving technique with a hand-tensioned loom (fig. 7.4). The shape of the wooden stand that the weaver is using resembles a foot. The weaving technique is weft-rib weave. Sandals are woven while the weaver is pulling the end of the warp with the left hand. The material for both warp and weft is rice straw. This picture was taken at Hachijo-jima, Tokyo, in 1987.

7.2.3 Foot-Tensioned Loom

This picture is of *Mokko* weaving with a foot-tensioned loom (fig. 7.5), taken at Taketomi-jima, Okinawa prefecture in 2011. *Mokko* is a type of woven basket which was formerly used as a carrier in construction sites, and such *Mokko* weaving had been done until the 1970s. Tension in the warp is applied by stepping on both ends of the warp with the feet. The weaving technique is weft-twined weave, carried out by both hands. The material for both warp and weft is rice straw.

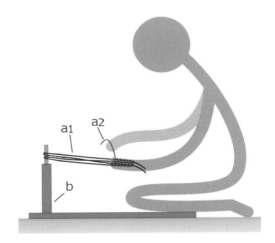

Fig. 7.4 Sandal weaving with a hand-tensioned loom; diagram: a1—warp, a2—weft, b—stand.

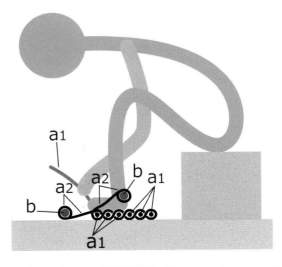

Fig. 7.5 Foot-tensioned loom. The National Museum of Ethnology, Osaka, H0270196; diagram: a1—warp, a2—weft, b—stick of coral for winding weft.

A World of Looms: Weaving Technology and Textile Arts

 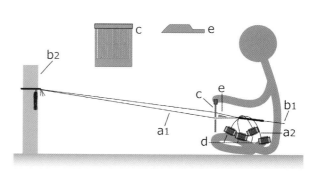

Fig. 7.6 Back-tensioned loom; diagram: a1–warp, a2–weft, b1–back strap, b2–pillar, c–plate-heddle with holes and slots, d–spool (4 pieces), e–weft beater.

 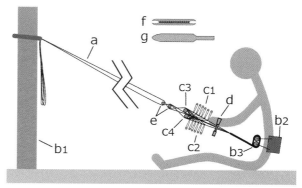

Fig. 7.7 Back-tensioned loom; diagram: a–warp, b1–pillar, b2–back strap, b3–cloth beam, c1–loops-heddles (6 pieces), c2–loops-heddles (6 pieces), c3–shed stick, c4–shed stick, d–reed, e–lease rods, f–spool, g–weft beater.

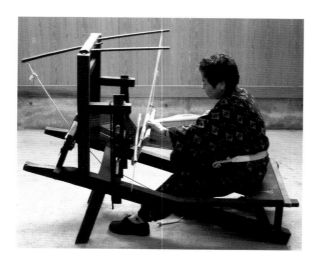 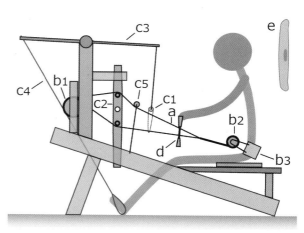

Fig. 7.8 Back-tensioned loom with a half-frame. The National Museum of Ethnology, Osaka, H0229422; diagram: a–warp, b1–warp beam, b2–cloth beam, b3–back strap, c1–loops-heddle, c2–shed frame, c3–balancing arm rod, c4–pulling cord, c5–presser rod for upper warp, d–reed, e–combined weft beater and spool.

7.2.4 Back-Tensioned Loom (1)

This picture is of band (*Batou-obi*) weaving with a back-tensioned loom (fig. 7.6), which was done at the Tsunan Museum of History and Folklore in 2003. A plate-heddle with holes and slots is used as the shedding device. This type of loom was used exclusively for *Batou-obi* weaving at Tsunan, Niigata prefecture and in the surrounding areas. The weaving technique is warp-rib weave. The material of both warp and weft of *Batou-obi* is *Boehmeria nivea* (ramie).

7.2.5 Back-Tensioned Loom (2)

This picture is of band weaving called *Kappeta-ori* with a back-tensioned loom (fig. 7.7). The band is a type of patterned fabric with double warp-rib weave. For the shedding device, twelve loops-heddles and two shed sticks are used. The material of both warp and weft is cotton. This picture was taken at Hachijo-jima, Tokyo in 1987.

7.2.6 Back-Tensioned Loom with a Half-Frame (1)

This picture, taken at Aomori in 2003, is of rug weaving with a more complex type of back-tensioned loom, with a "half-frame." (fig. 7.8) The large, paddle-shaped warp beam is lodged behind two uprights in the loom frame, while the cloth beam is attached to the weaver's waist and remains "free floating," as with the back-tensioned looms described above. A loops-heddle and a shed stick are used as the shedding devices. The counter shed of the warp is made with the loops-heddle which is pulled up by a cord attached to the weaver's right foot. The material of warp and weft is cotton. The warp is cotton yarn, but the weft is made of old recycled clothes torn to thin strips. The loom is of a similar type to the back-tensioned loom in R. O. Korea (section 7.4.2), and also resembles some looms that are found in southern China (section 6.3).

7.2.7 Back-Tensioned Loom with a Half Frame (2)

This is another example of a back-tensioned loom with a half-frame (fig. 7.9), of which there are many variants in Japan. This particular loom was used to weave gauze cloth. As shedding devices, two loops-heddles and one shed rod are used. The two loops-heddles are controlled by the right and left feet. The weaver alternately draws the pulling-cords by each foot for pulling up the two loops-heddles that are necessary to construct a gauze weave. The material of both warp and weft is *Boehmeria nivea* var. *nipononivea* (ramie). This loom was used until around the 1970s at Muikamachi, Niigata prefecture.

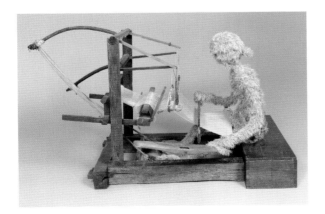
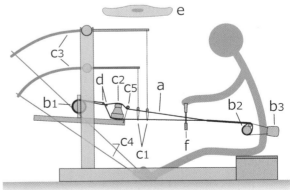

Fig. 7.9 Back-tensioned loom with a half-frame. The National Museum of Ethnology, Osaka, H0270363; diagram: a—warp, b1—warp beam, b2—cloth beam, b3—back strap, c1—loops-heddle (2 pieces), c2—shed rod, c3—balancing arm rod (2 pieces), c4—pulling cord, c5—presser rod for upper warp, d—lease rods, e—combined weft beater and spool, f—reed.

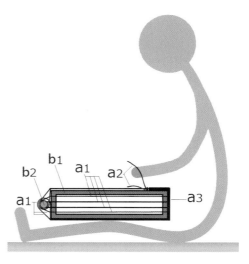

Fig. 7.10 Core-tensioned loom. National Museum of Ethnology, Osaka, H0270182; diagram: a1–warp, a2–weft, a3–woven fabric (part of a woven bag), b1–core (wooden box), b2–supporting stick for false circular warp.

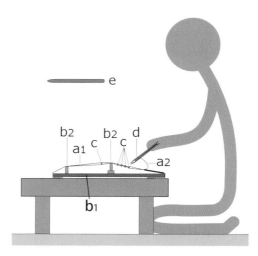

Fig. 7.11 Board-tensioned loom; diagram: a1–warp, a2–weft, b1–board, b2–warp tension adjusting sticks (3 pieces), c–pattern sticks (5 pieces), d–spool, e–weft beater.

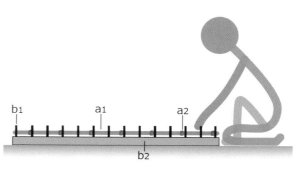

Fig. 7.12 Rod-tensioned loom; diagram: a1–warp, a2–weft, b1–rods (52 pieces), b2–wooden frame.

7.2.8 Core-Tensioned Loom

This picture is of shoulder bag weaving with a core-tensioned loom (fig. 7.10), meaning that the warp is tensioned by winding it tightly around a core, in this case a wooden box. This particular shoulder bag was woven with weft-twined weave, for which no shedding devices were used, the warp and weft manipulated entirely by hand. The warping system on this type of loom is false circular warp, with a supporting stick. The warp is wound around the core and reverses direction around the supporting stick, so that the shoulder bag is released when the stick is removed. The material of warp and weft is *Carex morrowii* (Boott), commonly known as Japanese sedge. This shoulder bag weaving was done at Nakanojo, Gunma prefecture in 2011.

7.2.9 Board-Tensioned Loom

This is a board-tensioned loom (fig. 7.11). This loom has been used to weave patterned fabrics that are called *Saga-nishiki* at Saga from the early 19th century onwards. The weaving technique is a type of patterned weave. This loom has no heddle, but pattern sticks are used. The material of the warp is gold leaf pasted on Japanese paper, with colorful silk threads used as weft. The warping system of this loom is flat warp. The tension in the warp is applied with the wooden board aided by three sticks. The ends of the warps are glued to the rear of the board.

7.2.10 Rod-Tensioned Loom

This picture shows rice-straw net weaving with a rod-tensioned loom (fig. 7.12). In this weaving the warp is stretched over the rods on the wooden frame. This type of rice straw net is used as a support for raising silkworms. The net is woven with a weft-twined weave: no tools are used and the net is woven entirely by hand. This net weaving was done at Nakanojo, Gunma prefecture in 2011.

7.2.11 Hanging-Weight-Tensioned Loom

This picture shows *Mokko* weaving with a hanging-weight tensioned loom (fig. 7.13). The technique is warp twined weave, carried out by both hands. The material for warp and weft is *Alpinia zerumbet* (shell ginger). The *Mokko* was woven at Taketomi-jima, Okinawa prefecture in 2011.

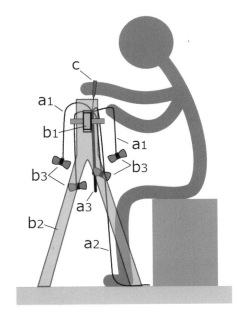

Fig. 7.13 Hanging-weight tensioned loom. The National Museum of Ethnology, Osaka, H0270202; diagram: a1–warp, a2–weft, a3–part of a woven *Mokko*, b1–beam, b2–stand, b3–wooden weights for spool (12 pieces), c–needle.

A World of Looms: Weaving Technology and Textile Arts

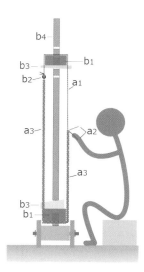

Fig. 7.14 Vertical frame-tensioned loom; diagram: a1–warp, a2–weft, a3–part of a woven mat, b1–warp beam, b2–supporting stick for knotted circular warp, b3–wedge, b4–side pillar.

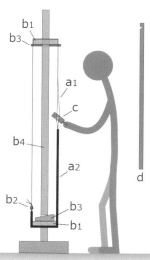

Fig. 7.15 Vertical frame-tensioned loom. The Tsunan History and Folklore Museum, Niigata prefecture, 1753; diagram: a1–warp, a2–part of a woven mat, b1–warp beam, b2–supporting stick for knotted circular warp, b3–wedge, b4–side pillar, c–bar heddle with triangular slots, d–weft insertion stick.

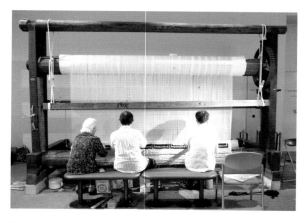
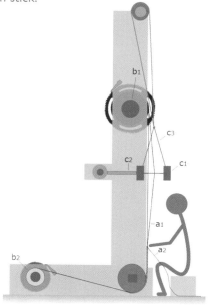

Fig. 7.16 Vertical frame-tensioned loom. The Sakai City Museum, Osaka; diagram: a1–warp, a2–weft, b1–warp beam, b2–cloth beam, c1–frame heddle with holes and slots, c2–prop, c3–hanging rope.

7. Japanese, Ainu, and Korean Looms

7.2.12 Vertical Frame-Tensioned Loom (1)

This picture shows mat weaving with a vertical frame-tensioned loom (fig. 7.14). The mat is woven with the weft-twined weave technique: no shedding devices are used and both hands are used to weave. The warping system of this loom is knotted circular warp, and a supporting stick is used. The material for both warp and weft is rice straw. This particular mat weaving was done at Nakanojo, Gunma prefecture in 2011.

7.2.13 Vertical Frame-Tensioned Loom (2)

This is another type of vertical frame-tensioned loom (fig. 7.15), also used for making rice straw mats. In this case the weaving technique is plain weave. The warping system is knotted circular warp and a supporting stick is used. A bar-heddle with triangular slots is used as the shedding device, and this heddle is also used as weft beater. This loom is similar to the frame-tensioned loom in R. O. Korea that is also used to weave rice straw mats (section 7.4.3).

7.2.14 Vertical Frame-Tensioned Loom (3)

This picture shows *Sakai-dantsu* weaving with a vertical frame-tensioned loom (fig. 7.16). *Sakai-dantsu* is a pile carpet that has been woven from 1831 onwards at Sakai and its surroundings, Osaka prefecture. The weaving technique is plain weave for the ground weave. A frame-heddle with holes and slots is used as the shedding device: this is probably the widest in the world of its type. The material for warp, weft and pile are cotton.

7.2.15 Inclined Frame-Tensioned Loom

This type of inclined frame-tensioned loom (fig. 7.17) was in use until the 1930s. Cloth was made with recycled materials on this loom, and with plain weave. A loops-heddle is used to make the counter-shed, while a bar attached to the frame is used to maintain the natural shed. These are controlled by each of the treadles which are operated by the weaver's right and left feet. The overall form of this loom, with its central post supporting the heddle lifting mechanism, has some similarities with the Han Dynasty looms on the Asian mainland (cat. 8). The material of both warp and weft is cotton, with recycled clothes torn into thin strips used as weft.

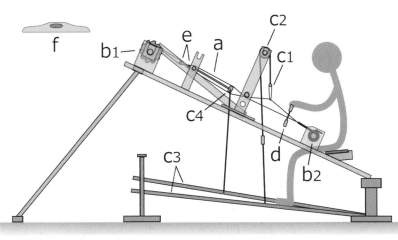

Fig. 7.17 Inclined frame-tensioned loom. Hakusan-Studio at Shiramine, Hakusan, Ishikawa prefecture; diagram: a–warp, b1–warp beam, b2–cloth beam, c1–loops-heddle, c2–pulley, c3–treadles, c4–shed frame, d–reed, e–lease rods, f–combined weft beater and spool.

A World of Looms: Weaving Technology and Textile Arts

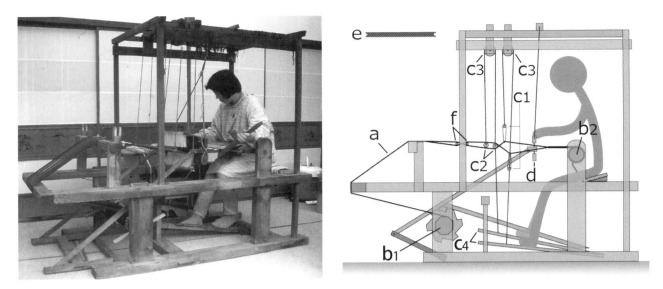

Fig. 7.18 Frame-tensioned loom. The National Museum of Ethnology, Osaka, H0216594; diagram: a–warp, b1–warp beam, b2–cloth beam, c1–double loops-heddles, c2–double shed sticks, c3–pulley, c4–pair of treadles, d–reed, e–spool, f–lease rods.

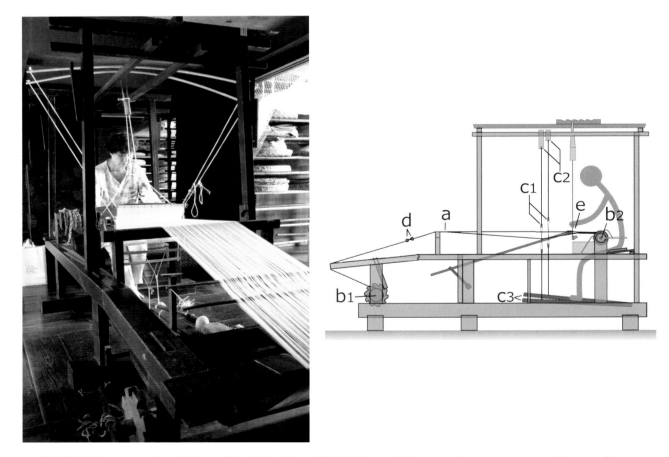

Fig. 7.19 Frame-tensioned loom. The Fukushima City Minka-en Museum; diagram: a–warp, b1–warp beam, b2–cloth beam, c1–jointed loops-heddles (2 pieces), c2–curved bamboo springs (2 pieces), c3–pair of treadles, d–lease rods, e–reed.

7. Japanese, Ainu, and Korean Looms

7.2.16 Frame-Tensioned Loom (1)

This loom was used to weave mats with plain weave (fig. 7.18). Double loops-heddles and two shed sticks are used as shedding devices. These are controlled by treadles: the treadle for the double loops-heddles is operated by the right foot and the treadle for the double shed stick is operated by the left foot. The material of the warp is cotton and the weft consists of old clothes torn to thin strips. This particular mat weaving was done at Haramura, Nagano prefecture in 1999.

7.2.17 Frame-Tensioned Loom (2)

This picture is a demonstration of silk weaving with a restored frame loom (fig. 7.19) that was used in Fukushima city until the 1980s. The weaving technique is plain weave. Two jointed loops-heddles are used as shedding devices: these are attached to the ends of curved bamboo springs, in a similar manner to the ground-weave heddles of Chinese drawlooms (section 5). These are attached to treadles operated by the weaver's right and left feet. This type of loom was a general loom for silk weaving in Japan during the Edo Period (1603–1868).

7.2.18 Frame-Tensioned Loom (3)

This picture is of a frame-tensioned loom (fig. 7.20) that was formerly used to weave silk weft ikat at Shiozawa, Niigata prefecture until the 1970s. The weft ikat was woven with plain weave. Double jointed loops-heddles were used as shedding devices.

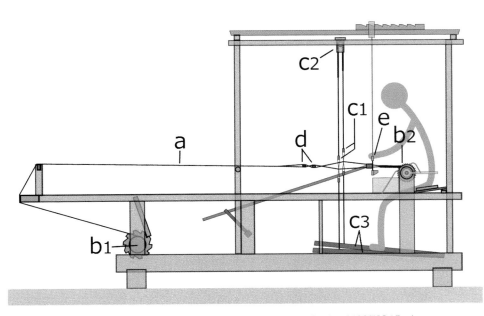

Fig. 7.20 Frame-tensioned loom. The National Museum of Ethnology, Osaka, H0270365; diagram: a–warp, b1–warp beam, b2–cloth beam, c1–double jointed loops-heddles, c2–roller, c3–pair of treadles, d–lease rods, e–reed.

A World of Looms: Weaving Technology and Textile Arts

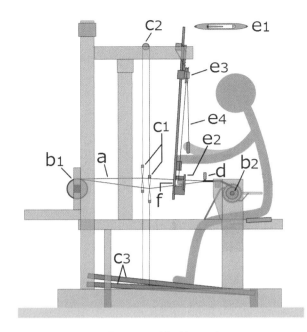

Fig. 7.21 Frame-tensioned loom. The National Museum of Ethnology, Osaka, H0270366; diagram: a–warp, b1–warp beam, b2–cloth beam, c1–double jointed loops-heddles, c2–roller, c3–pair of treadles, d–tenter, e1–shuttle, e2–rail box for flying shuttle, e3–pulley, e4–release cord, f–reed.

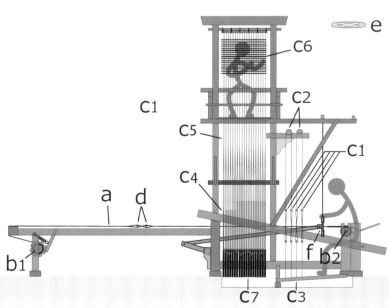

Fig. 7.22 Frame-tensioned loom (Drawloom). National Museum of Ethnology, Osaka, H0236950; diagram: a–warp, b1–warp beam, b2–cloth beam, c1–double false jointed loops-heddles, c2–roller (2 pieces), c3–treadles (4 pieces), c4–harnesses, c5–draw cords, c6–pattern cords, c7–weights, d–lease rods, e–shuttle, f–reed.

7.2.19 Frame-Tensioned Loom (4)

This type of frame-tensioned loom (fig. 7.21) was used to weave silk weft ikat with a plain weave structure, at Tamamura, Gunma prefecture until the 1970s. Double jointed loops-heddles were used as shedding devices. The loom is equipped with a reed-beater attached to a swinging bar, and a European-style flying shuttle.

7.2.20 Drawloom

This is a drawloom (fig. 7.22), a type of frame-tensioned loom used to weave patterned fabric, in which patterns are stored above the loom for use and re-use. This particular loom was used in Tokyo in the late 20th century. In this loom, two pairs of double false jointed loops-heddles are used as the shedding devices for the ground weave. Harnesses, draw strings and pattern cord are used as shedding devices for patterning: these are operated by a weaver's assistant (draw person), who sits in the tower above the loom. The general form of this loom is similar to Chinese drawlooms of the type called the Lesser Drawloom, such as the damask loom (cat. 20).

7.3 Ainu Looms

7.3.1 Hand-Tensioned Loom

This picture shows a hand-tensioned loom used to weave a sword-hanging band (fig. 7.23), with weft-twining. This weaving is limited to the center part of the warp. Tension is applied to the warp by hand, while both hands are used to do the weft-twined weave. The material for warp is *Tilia japonica* (a type of linden) and the weft is cotton. This demonstration of band weaving was done at the Ainu Museum, Shiraoi, Hokkaido in 2012.

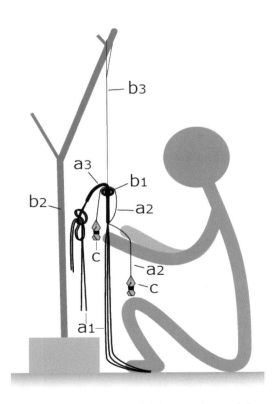

Fig. 7.23 Hand-tensioned loom; diagram: a1—warp, a2—weft, a3—part of a woven band, b1—warp beam, b2—tree branch, b3—hanging cord, c—reel.

7.3.2 Hanging-Weight-Tensioned Loom

This shows a hanging-weight tensioned loom (fig. 7.24), similar to the loom described above for weaving *Mokko* (fig. 7.13). This particular loom was made at Nibutani, Biratori, Hokkaido in 1978. The weaving technique of the mat is warp-twining, done by hand. The material for warp is *Tilia japonica* and the weft is *Typha latifolia* L.

7.3.3 Back-Tensioned Loom (1)

This is a back-tensioned loom (fig. 7.25) which was used at Noboribetsu, Hokkaido to weave narrow bands prior to 1936. A loops-heddle and a shed board are used as shedding devices. The shed board serves to hold open the natural shed opening, and also works as a warp spacer. Both these shedding devices are controlled by hand. The material of the warp is cotton and the weft is *Tilia japonica*.

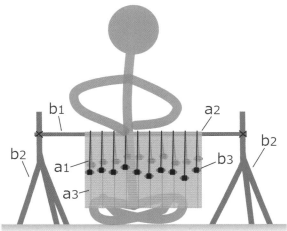

Fig. 7.24 Hanging-weight tensioned loom. The National Museum of Ethnology, Osaka, H0062481; diagram: a1—warp, a2—weft, a3—part of a woven mat, b1—bar, b2—stand, b3—weights (48 pieces).

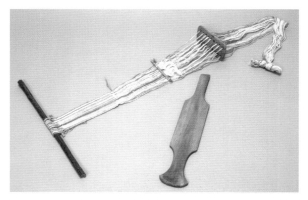
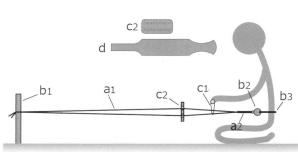

Fig. 7.25 Back-tensioned loom. The National Museum of Ethnology, Osaka, H0018709 and H0018710; diagram: a1—warp, a2—woven cloth, b1—stake, b2—cloth beam, b3—back strap, c1—loops-heddle, c2—shed board and warp spacer, d—weft beater.

7.3.4 Back-Tensioned Loom (2)

Traditionally headbands and sword hanging bands had been woven by the Ainu using hand-tensioned looms (fig. 7.26, section 7.3.1). But both these textiles have been woven using back-tensioned looms at Nibutani, Hokkaido since the 1960s. The reason is that the back-tensioned loom can weave the weft-twined textiles more easily than the hand-tensioned loom and the back-tensioned loom does not require the weaver to control the tension in the warp by hand. The picture shows a headband weave using the back-tensioned loom at Nibutani in 2016. The weaving, using weft-twining, is limited only to the center part of the warp and is done using both hands. The material for warp and weft is *Ulmus laciniata* (a type of elm). This headband weaving and also sword hanging band weaving by the back-tensioned loom are examples of "modernization" of traditional weaving techniques amongst the Ainu.

7.3.5 Back-Tensioned Loom (3)

This is a back-tensioned loom (fig. 7.27) that was used to make plain weave textiles at Shizunai, Hokkaido until 1937. A loops-heddle and a shed stick are used as the shedding devices, both of which are worked by hand. A reed is also used in this loom, but this is not used as a beater and only functions as a warp spacer. The material of the warp and weft is *Tilia japonica*.

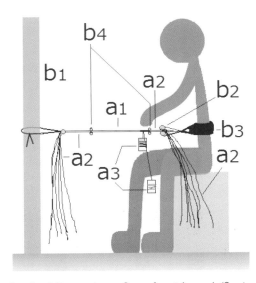

Fig. 7.26 Back-tensioned loom; diagram: a1—warp, a2—warp for braiding string, a3—weft with reel (2 pieces), b1—pillar, b2—cloth beam, b3—back strap, b4—warp fixing sticks.

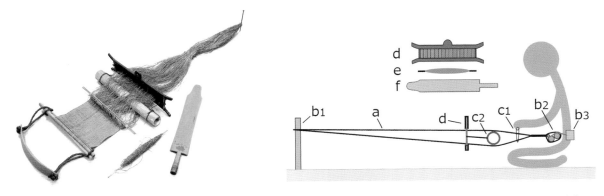

Fig. 7.27 Back-tensioned loom. The National Museum of Ethnology, Osaka, H0011021; diagram: a—warp, b1—warp beam, b2—cloth beam, b3—back strap, c1—loops-heddle, c2—shed rod, d—reed, e—spool, f—weft beater.

7.4 Looms on the Korean Peninsula

It is recorded in the *Nihon-shoki* (the oldest chronicles of Japan) that a brocade weaver came to Japan from Baekje in 463. At that time Baekje was located in the midwest of the Korean Peninsula. After the time of the *Nihon-shoki*, many names of brocades and other patterned fabrics appear in the historical records of the Korean Peninsula. Therefore, there is no doubt that various patterned fabrics were woven on the Korean Peninsula, though it is unclear what types of looms were used to weave them. There are also many questions remaining about other types of fabrics and their looms.

On the Korean Peninsula the following three looms are used, though this is not a complete list. In addition, some of the same types of the Japanese body-tensioned loom used for sandal weaving (section 7.2.1), the hanging-weight tensioned loom for warp-twined weave (section 7.2.11) and the vertical frame-tensioned loom for weft-twined weave (section 7.2.12), are also used on the Korean Peninsula.

7.4.1 Back-Tensioned Loom

This is a back-tensioned loom (fig. 7.28) that was exhibited at the Buyeo National Museum, R. O. Korea, in 2008, used to weave sandals. The weaving technique is weft-rib weave. For this weft-rib weave, no shedding devices are used and the sandal is woven using only the hands. The material for warp and weft is rice straw.

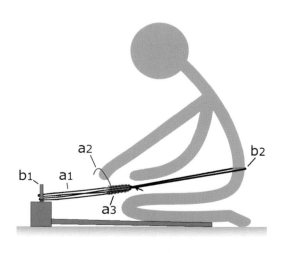

Fig. 7.28 Back-tensioned loom; diagram: a1–warp, a2–weft, a3–part of a woven sandal, b1–stand, b2–back strap.

7.4.2 Back-Tensioned Loom with a Half-Frame

This is ramie cloth weaving using a back-tensioned loom (fig. 7.29), used in Buyeo of R. O. Korea. An example of this loom is included in the catalog (cat. 24). The weaving technique of this loom is plain weave and a loops-heddle and shed stick are used shedding devices. The counter-shed is made by the heddle which is pulled up by the weaver's right foot. This loom is the same basic type as a Japanese back-tensioned loom (section 7.2.6).

7.4.3 Vertical Frame-Tensioned Loom

This is a vertical frame-tensioned loom (fig. 7.30) used for weaving a rice straw mat. The weaving technique of this mat is weft-rib weave and the warping system is knotted circular warp, using a supporting stick. A bar-heddle with triangular slots is used as the shedding device and this bar-heddle is also used as a weft beater. This particular loom was used at Andong-gun, Gyeongsangbuk-do, R. O. Korea, until 1980. This loom is a similar type to the Japanese vertical frame-tensioned loom to weave rice straw mats (section 7.2.13).

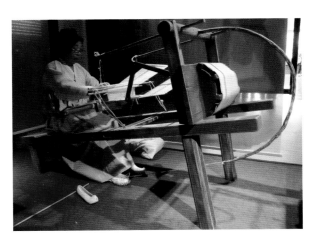
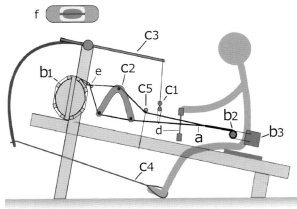

Fig. 7.29 Back-tensioned loom; diagram: a–warp, b1–warp beam, b2–cloth beam, b3–back strap, c1–loops-heddle, c2–shed sticks, c3–balancing arm rod, c4–pulling cord, c5–warp depression rod, d–reed, e–lease rods, f–shuttle.

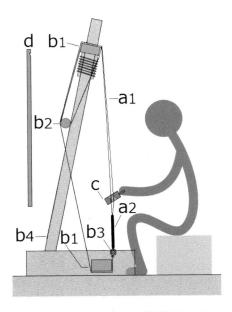

Fig. 7.30 Vertical frame-tensioned loom. The National Museum of Ethnology, Osaka, H0085449; diagram: a1– warp, a2–part of a woven mat, b1–warp beam, b2– warp tension adjusting rod, b3–supporting stick for false circular warp, b4–side pillar, c–bar heddle with triangular slots, d– weft insertion stick.

8. Voice of the Weaver: Hangzhou

Yu Youde works at the China National Silk Museum as a weaving demonstrator. He apprenticed under Luo Qun (fig. 8.1).

I live in the suburbs near Hangzhou, I had been a farmer until I was hired by the China National Silk Museum as a gardener. I just studied one year in elementary school, so I did not have the confidence to learn weaving skills initially. The [museums'] staff encouraged me. All the weaving skills I have, I learned from Luo [Qun]… I call him master. My job is to demonstrate weaving processes to visitors generally. I can weave silk—the kinds seen in ancient China, such as jin *silk, brocade, patterned damask, patterned gauze, [and] I can set up a loom. Younger people don't like to do this job; in fact, the job is interesting. Solving problems one by one during loom set-up is a pleasure. I feel useful in protecting intangible cultural heritage.*

Fig. 8.1 Yu Youde (right) and Luo Qun (photo: Luo Qun).

Fig. 8.2 Laoguanshan reconstruction loom (photo: Luo Qun).

Luo Qun is a senior researcher and master weaver at the China National Silk Museum. He was in charge of the team that reconstructed the ancient Laoguanshan loom. He believes that the Laoguanshan loom (fig. 8.2, cat. 7) is different from other looms in ancient China (fig. 8.3).

In 2013, four wooden pattern loom models were excavated from a Han Dynasty burial in Laoguanshan, Chengdu, Southwest China. In technical terms, these looms should be called pattern looms with multi-shafts and moving hooks. They were the most developed pattern looms in the world at their time… All the information of the pattern were stored in series of shafts…[This system is] different than the drawloom developed in the Ming and Qing Dynasties, which uses drawstrings to store pattern information. The Laoguanshan loom uses complex mechanism to move one pair of hooks and to lift shafts one by one…

We have reconstructed a life-size loom according to the loom models. I cooperated with other weavers to fasten the heddles on the shafts; one shaft has 2,000 heddles; 84 shafts have to be fastened; and 10,470 warp ends have to be passed through the heddle loops or the gaps between heddles. The work took more than one and a half years. Now we have reproduced the famous jin *silk with the Chinese characters* 五星出东方利中国 *, which was excavated from the site of "Jing Jue," the ancient kingdom in southern Xinjiang Uygur autonomous region, dating back to the Eastern Han Dynasty.*

Fig. 8.3 Luo Qun demonstrating weaving on the Laoguanshan reconstruction loom (photo: Christopher Buckley).

Southeast Asia

9. Mainland Southeast Asian Looms
   Linda McIntosh

10. Insular Southeast Asian Looms
    Sandra Sardjono

11. Voice of the Weaver: Palembang and Muang Kham
    Zainal Arifin, Bouakham Phengmixay

9. Mainland Southeast Asian Looms

Linda McIntosh

Cambodia, the Lao People's Democratic Republic (commonly known as Laos), Myanmar (formerly Burma), Thailand, the Socialist Republic of Vietnam, and Peninsular Malaysia make up Mainland Southeast Asia. Modern geo-political borders divide settlements of the same or closely related ethnic groups; thus, related cultural traditions, including textiles and knowledge regarding textile production, exist in different countries throughout the region. The majority of the region's population practiced and many ethnic groups continue to practice subsistence agriculture as a primary occupation. Exceptions include peoples living in coastal areas of South Vietnam and Cambodia and Peninsular Malaysia where regional maritime commerce is more important than farming.

The ethnic groups in Mainland Southeast Asia are diverse, belonging to the Austronesian, Austroasiatic, Hmong-Mien, Sino-Tibetan, and Tai-Kadai language families. Weavers in this region utilize a variety of looms to produce cloth, from foot-braced body-tensioned looms to large looms with rigid frames. The range of looms found in Mainland Southeast Asia reflects the region's cultural and historical links to other areas, especially Southwest China (section 6). Exposure and interaction between various cultures via trade, hegemony, marital, and other alliances have affected the spread and assimilation of weaving technology including looms. During his visit to the Angkor Empire of Cambodia at the end of the 13th century, Zhou Daguan, a Chinese diplomat, observed that the Khmer wove cotton on backstrap looms, while international trade provided textiles used by the court, and Siamese (Thai) weavers wove silk textiles on rigid frame looms.[1] Khmer women assimilated the technique of weaving silk on frame looms to the extent that production was both extensive and commercialized by the 18th century. The Central Thai kingdom of Ayutthaya imported weft-ikat decorated silk textiles from Cambodia on a regular basis.[2] As a result of this complex history, looms and knowledge related to textile production can vary widely between subgroups of branches in the same ethnolinguistic family. For example, many Khmu subgroups of the Austroasiatic ethnolinguistic family do not weave, while a few produce cloth on body-tensioned looms.

The roles of textiles in Southeast Asian societies parallel their functions in other regions such as China and Insular Southeast Asia. Handwoven fabrics serve as markers of ethnicity and status including marital status and wealth. Women are traditionally the primary producers of handwoven cloth, and society once evaluated a woman's worth as a good member of the community or a diligent and patient wife and mother according to her weaving skills.[3] Knowledge regarding textile production was passed down from one generation of women to another, and adolescent girls began preparing their own trousseaus by weaving cloth that would serve as clothing, bedding, and other household accessories, and as gifts for her future relatives. After marriage, they continued to weave for members of their households and to create textiles to use as offerings and attire in both religious and secular contexts.

Facing page: detail of a Tai Daeng ceremonial cloth, made from silk warp and cotton weft, decorated with silk discontinuous supplementary weft. The geometric motifs (hooks and diamonds) and stylized animals (birds, naga) are characteristic of Southeast Asian textile designs.

9.1 Body-Tensioned Looms

The foot-braced body-tensioned loom (also known as a foot-braced backstrap loom), the most ancient type in the region, continues to be used by groups belonging to the Katuic and Bahnaric sub-branches, Khmer branch, of the Austroasiatic ethnolinguistic family, such as the Bahnar, Katu, Krieng, Mnong, and Ta-oi residing in Northeast Cambodia, South Laos, and Central and South Vietnam. Weavers from the O'du group, Khmuic sub-branch of the same language family that live in adjacent areas of Northeast Laos and Northwest Vietnam, reserve this loom type for producing cloth that will be made into shoulder bags (fig. 9.1). Cotton and synthetic threads, such as polyester, are the primary fibers woven on this type of loom. Generally, the weave is warp-faced, meaning that the warp threads are visible on a fabric's surface while the weft is hidden.

The decorative techniques used on cloth produced on a foot-braced backstrap loom in Mainland Southeast Asia include warp-float or complementary-warp patterning and weft twining (fig. 9.2). A distinctive method of forming designs is by inserting beads during weaving.[4] Katu and Ta-oi weavers string beads on the weft threads and strategically place beads between warp threads while inserting a weft in a shed to compose motifs. These ethnic groups do not produce the beads themselves but acquire them via trade and consider metal ones to be the most valuable, followed by glass. Presently, metal beads are not available in

Fig. 9.1 A weaver from the O'du ethnic group uses a foot-braced backstrap loom to weave cloth used to make shoulder bags. Khaap village, Khoun district, Xieng Khouang province, Laos, 2017 (photo: Linda McIntosh).

Fig. 9.2 A weaver from the Katu ethnic group applies warp-float or complementary-warp technique while weaving fabric for a tubular skirt. Huaysay village, Thataeng district, Sekong province, Laos, 2013 (photo: Linda McIntosh).

the market and are only found on older textiles, while plastic versions are making inroads as an inexpensive alternative to glass beads.

One panel of woven cloth may be sufficient for a tubular skirt that extends as far as the knees, but a couple of panels are often joined at the selvedges and warp ends to form a longer tube skirt. A woman could then tuck the skirt's top end, folded or unfolded, around her breasts. The long tube-skirt can also serve as a sleeping bag that can be pulled up over the head. Two or more lengths of fabric can be sewn together along the selvedges to create a shawl or blanket-like garment. Men traditionally wore a loincloth (fig. 9.3) consisting of a long narrow fabric several meters long that is folded around the waist and groin, and its use continues during special occasions by some males. Part of this garment also crisscrosses the torso when worn. Sleeveless tunics served as women's traditional upper garments in the past but have now been replaced with factory-made blouses.

The distribution of the externally braced body-tensioned loom with the warp beam secured to a stationary object such as a fence or between two posts is widespread in Mainland Southeast Asia. It is not found in Peninsular Malaysia but in the Malaysian part of Borneo, which is part of Insular Southeast Asia. Women that weave on this type of body-tensioned loom belong to all the ethnolinguistic families previously mentioned, with the exception of the Tai-Kadai. The distribution of this device ranges from Northeast Cambodia and South Vietnam, the homeland of the Jarai who speak an Austronesian language and the Bahnar of the Austroasiatic family (fig. 9.4), to the north and northwest corners of Myanmar, continuing into the Himalayan region. The Lawa of the Palaungic sub-branch, Austroasiatic family, live in North Thailand and related groups, the Wa of Myanmar and Palaung of both countries, also produce cloth on this kind of backstrap loom (fig. 9.5). Members of the Tibeto-Burman ethnolinguistic family that use versions of this loom type include Chin, Kachin, Naga, and Karen subgroups. The Karen groups are found in Thailand and Myanmar while the subgroups of the Chin and Kachin range over North Myanmar and neighboring parts of India and China.

Fig. 9.3 Loincloth made by the Katu people, cotton with warp-float patterning and weft twining, and small glass beads at the ends of the cloth, fringes made from wool, 38 cm x 400 cm. Private collection.

Fig. 9.5 While weaving, Palaung women wrap the warp around a horizontal warp beam fixed to a house-beam and keep the tension taut with the broad strap around their backs. Chiang Dao district, Chiang Mai province, Thailand, 2004 (photo: Linda McIntosh).

9.2 Looms with a Reed and a Frame

The inclusion of a reed increases the types of materials woven and decorative techniques that can be used in both body-tensioned and fixed cloth-beam looms, for example making balanced and weft-faced weaves possible. The reed also separates the warp threads preventing tangles. Silk and wool join cotton and synthetic threads as fibers woven on this loom type. Haka Chin weavers of Northwest Myanmar produce fabric with a silk warp and weft on this kind of loom. Textile producers of other ethnicities incorporate silk as weft and supplementary weft or brocade. Some Kachin subgroups use wool in the weft. Besides weft twining and complementary-warp weave, supplementary-warp and supplementary-weft techniques are also applied to create complex patterns while using a backstrap loom with a raised seat and frame in Mainland Southeast Asia.

Body-tensioned frame looms with a reed occur among ethnic minority groups with populations in China and adjacent areas of Mainland Southeast Asia (section 6.2). The number of Hmong (or Miao as this group is called in China) women who continue to weave in Laos, Thailand, and Vietnam has dwindled, but some living in Northeast Laos and North Vietnam continue to produce hemp textiles on a frame loom that employs a backstrap to secure the cloth beam at the weaver's waist and to control warp tension.

Fig. 9.4 Facing page: detail of cloth woven on an externally-braced backstrap loom by an Austroasiatic weaver from southern Vietnam, possibly Jarai group. Warp patterning and warp ikat dashes. Private collection.

9.3 Looms with Frames and Fixed Cloth Beams

Fig. 9.6 An Akha woman stands while weaving unpatterned, plain weave cloth. Sing district, Luang Nam Tha province, Laos, 2006 (photo: Linda McIntosh).

Some Akha (called Hani in China) women use a frame loom composed of rigid pillars including a post in the ground to stretch the warp taut and a cloth beam affixed to the opposite end of the frame. A weaver stands at the cloth beam to step on a pair of treadles connected to the principal heddles that open sheds for the insertion of weft thread (figs. 6.9 and 9.6). A long, narrow roll of plain woven cloth is produced.

The frame loom or loom composed of a rigid frame and affixed cloth and warp beams is used by groups from all of the language families found in Mainland Southeast Asia (section 6.5). Variations of this loom type are found throughout the region, but the main parts are similar. Posts at four corners of a rectangle are perpendicular to the ground, and beams that are parallel to the ground are attached to the posts to form a rigid frame. A weaver sits at one end in front of the cloth beam while the warp beam is fixed to the posts at the opposite end.

The warp beam can extend beyond these posts by extending the parallel beams on some looms (fig. 9.7). The reed-beater and heddles hang from cords tied to wood or bamboo poles lying across the top of the frame. The methods of storing warp threads on a loom also vary. Khmer and Cham weavers in Cambodia and Lao textile producers in South Laos wrap the threads around a flat plank of wood while women belonging to groups from the Tai branch of the Tai-Kadai ethnolinguistic family in Laos, Vietnam, Thailand, and Myanmar wind the warp on a round rod, or fold the warp back over the top of the loom and tie it off in a loose knot above the weaver's head. The plank-like warp beam on Malay frame looms rests upright or vertical to the ground while the flat warp beam on Khmer looms is held in a horizontal position by cords (fig. 9.9).

Some loom parts are movable and interchangeable depending on the fabric to be produced. Reeds

9. Mainland Southeast Asian Looms

Fig. 9.7 On this Khmer loom in Northeast Thailand, the warp extends beyond the loom's vertical posts. Bau village, Sisaket province, Thailand, 2009 (photo: Linda McIntosh).

occur in different widths (from a few centimeters up to a meter), and a weaver selects a reed based on the cloth she intends to produce. The reed is enclosed in a frame, and it also serves as a beater. All looms with a rigid frames employ a pair of bidirectional heddles (clasped heddles) that separate even and odd-numbered warp threads. Each heddle is attached to a treadle. The arrangements of the treadles on the loom vary throughout the region: the treadle position is often perpendicular to the warp in Cambodia while it is parallel in Thailand. Generally, a balanced or 1:1 plain weave is produced for un-patterned fabric and for the ground weave of a decorated cloth. Textiles produced on rigid frame looms are used for many kinds of characteristic attire, including tube skirts, hip wrappers, shoulder cloths, head cloths and shawls, as well as household accessories and ritual items.

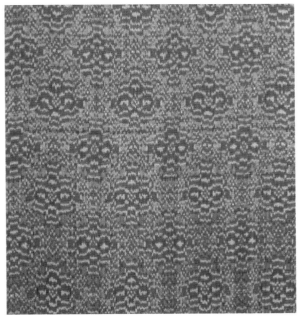

Fig. 9.8 Detail of a silk skirt from Cambodia in 2:1 twill weave with weft ikat patterning. Private collection.

73

 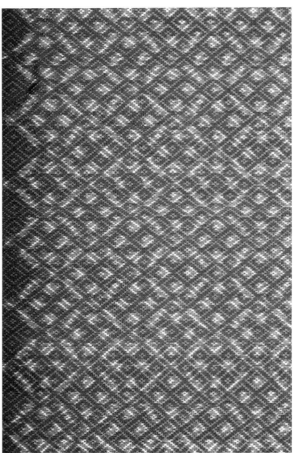

Fig. 9.9 The warp beam of a Khmer or Cambodian loom is flat and positioned parallel to the ground when secured to the loom's frame (photo: Linda McIntosh).

Fig. 9.10 Detail of the Iboeuk diamond twill design decorating a scarf by Artisans d'Angkor. Siem Reap, 2017 (photo: Linda McIntosh).

The type of weave and decorative technique(s) that a weaver selects influences the equipment that is mounted on a frame loom. A twill weave requires additional heddles with corresponding treadles. The weft ikat-decorated silk textiles of Cambodia are woven in a 2:1 twill weave (fig. 9.8), which requires three heddles/treadles to create. The diamond twill design called Iboeuk in Khmer language requires four heddles/treadles, and other twill weave patterns consist of up to six in this region (fig. 9.10). Khmer women in Northeast Thailand traditionally produce similar twill-patterned textiles while weavers from neighboring groups from Tai subgroups of the Tai-Kadai ethnolinguistic family in Laos and Thailand apply twill weave to cotton fabrics used as household accessories. Assimilation of the combination of weft ikat and twill techniques occurred near Inle Lake, southern Shan State, in present-day Myanmar in the early 20th century. During the British colonial period, weavers from the Intha ethnic group from the Tibeto-Burman branch, Sino-Tibetan language family, travelled to Northeast Thailand where weft ikat and twill adorned silk textiles in this region as well as in Cambodia. They returned to Myanmar and introduced these techniques into the local production of silk.[5]

Three methods of applying the supplementary warp technique on a frame loom occur in this region. In all three methods, a weaver installs the supplementary warp on the loom, inserting warp threads through additional heddles (fig. 9.11). There is one heddle per unique section of a design, and the supplementary heddles are placed behind

Fig. 9.11 The white supplementary warp threads are secured below the main warp. Phong ethnic group, Saleuy village, Sam Neua district, Hua Phan province, Laos, 2013 (photo: Linda McIntosh).

Fig. 9.12 A Katang weaver lowers heddles storing supplementary warp patterning by stepping on treadles. Vongsikeo village, Phine district, Savannakhet province, Laos, 2004 (photo: Linda McIntosh).

the main heddles. Weavers from Tai subgroups use a metal hook to lift a supplementary heddle and open a shed to form a pattern section. Katang and Mankong weavers from the Austroasiatic family use additional treadles to lower corresponding supplementary heddles (fig. 9.12).[6] Weavers from the lowland Eastern Cham ethnic group of the Malayo-Polynesian branch, Austronesian ethnolinguistic family in Vietnam apply supplementary warps using a distinctive loom using brass weights to form patterning on narrow cloth panels.[7]

Fig. 9.13 A Tai Yuan skirt showing hand-picked supplementary-weft designs in the border, silk supplementary-weft patterns on silk ground. Warps are oriented horizontally in this tubeskirt, 79 cm x 93 cm. Private collection.

9.4 Supplementary Weft and Other Decorative Techniques

Weavers also use different methods to apply continuous and discontinuous supplementary weft (brocade). Some, such as Tai Yuan weavers of Thailand, painstakingly pick out each line of a pattern by hand without the use of a device to store the pattern (fig. 9.13). Others insert sticks in the warp behind the main heddles to preserve the design for one-time reuse, which is particularly useful when making symmetrical designs. Each rod raises specific warp threads to form a shed for the insertion of a single weft shot comprising part of the supplementary weft design. A complex heddle consisting of a set of leashes supported above the warp allows supplementary patterns to be stored and applied as many times as desired (figs. 9.14–9.16, figs. 6.1 and 6.10). Each line of a pattern is recorded by a stick or string inserted into the leashes, and tens, hundreds, or thousands of strings can compose a design (section 6.6).

Women in Mainland Southeast Asia also apply other decorative techniques during the weaving process that do not require extra devices on the loom. One such technique is dovetail tapestry utilized by weavers belonging to Burman and some Tai subgroups. Ikat or resist-dye techniques are applied to warp or weft threads prior to weaving. Simple warp ikat designs of dashes occur on cloth produced by weavers belonging to Khmer and other groups from the Austroasiatic family and Tai subgroups of the Tai-Kadai languages. Complex patterns can be made using the weft ikat technique, and the distribution of this technique covers all nations and all ethnolinguistic families present in this region.

Fig. 9.14 Supplementary heddle leashes hang from the top of a loom. Phuthai ethnic group, Lahanam village, Songkhone district, Savannakhet province, Laos, 2004 (photo: Linda McIntosh).

Notes

1 Zhou 2007.
2 Baker 2011.
3 McIntosh 2008.
4 McIntosh 2014.
5 Fraser-Lu 1988, 97.
6 McIntosh 2009.
7 Howard 2008.

9. Mainland Southeast Asian Looms

Fig. 9.15 Blanket or bedcover made by Tai Dam weaver, with a repeating design made using a pattern stored on a vertical supplementary heddle. Indigo cotton supplementary weft on cotton ground, two strips of cloth sewn together, 80 cm x 130 cm. Private collection.

Fig. 9.16 Facing page: detail of a Tai Daeng ceremonial cloth, silk supplementary weft, silk warp and cotton weft. The presence of repeated "errors" and idiosyncrasies in the warp direction in this textile suggest the use of a pattern storage system such as the one in fig. 9.14. The cloth is about 30 cm wide. Private collection.

10. Insular Southeast Asian Looms

Sandra Sardjono

Insular Southeast Asia encompasses the modern nations of Indonesia, East Timor, the Philippines, Singapore, Brunei, and the part of Malaysia that is located in the northern Borneo. Weaving technology was first introduced into Insular Southeast Asia around 5000 BCE with the migrations of the Austronesian-language speaking peoples from southern China. Many scholars have theorized the routes of the Austronesian migration based on the studies of genetic, linguistic, and—most recently—by modeling loom evolution.[1] The Neolithic roots of these weaving traditions in Insular Southeast Asia are traceable in the widespread distribution of simple body-tensioned looms, the predominance of warp ikat technique, and the universal idea that textile production belongs to the realm of women.[2]

With the rise of organized polities in the last millennium BCE and the emergence of larger-scale statehoods and religions, the acquisition of power and prestige became a prime motivator for importing new ideas and technologies. The last two thousand years bear witness to the influx of foreign motifs and textile techniques from China, India, the countries of the Middle East and Europe into different parts of the region. The blending of these different traditions resulted in novel and creative ways of patterning seen in today's rich repertoire of Insular Southeast Asian weavings.[3] The looms used for weaving these various types of textiles can be broadly categorized into three main types based on the tensioning of the warp: body-tensioned looms, half-frame looms, and frame looms.[4]

Fig. 10.1 Tboli weaver at Lake Sebu, Mindanao, southern Philippines, 2016 (photo: Craig Diamond).

Facing page: detail of *abaca* cloth woven in Mindanao, southern Philippines. Warp stripes and warp ikat patterning. Private collection.

10.1 Body-Tensioned Looms with Circular Warps

The type of loom that came with the Austronesians is the portable, body-tensioned loom with continuous circular warp. Today, it is still the prevailing loom for weaving traditional textiles in the Lesser Sunda islands of Indonesia, north-central Sulawesi, northern Sumatra, Borneo, south-central Mindanao and northern Luzon. It is also found on the Asian mainland in Southeast Asia (section 9.1) and Southwest China (section 6.1). The main advantage of this loom is its relative ease of set-up. The weavers sit on the ground and use their bodies to manipulate the warp tension when changing the sheds. The cloth beam on one end of the loom is connected to the weaver's waist, while the warp beam on the other end is secured to a house post, a peg, or a ground frame (figs. 10.1 and 10.3, cat. 29–32). The externally braced, rather than the foot-braced system (where the weavers press their feet on the warp beam), is the norm in Insular Southeast Asia.[5]

Body-tensioned looms with circular warps are associated with the weaving of cotton yarns (fig. 10.2). In the past, bast and leaf fibers such as *abaca* were also regularly woven, but their use became less common during the last century. In the southern part of the Philippines, however, the

Above: fig.10.2 Sarong (*tais*) from Belu Regency, Timor, Indonesia. Woven on a loom similar to that in fig. 10.3 and made from three panels sewn together. The warp of each panel is horizontal in this photograph. Cotton with warp-ikat patterning and warp stripes, 55 cm x 130 cm. Private collection.

Right: fig.10.3 Weaver in Insana, Timor, 1973 (photo: Don Bierlich).

10. Insular Southeast Asian Looms

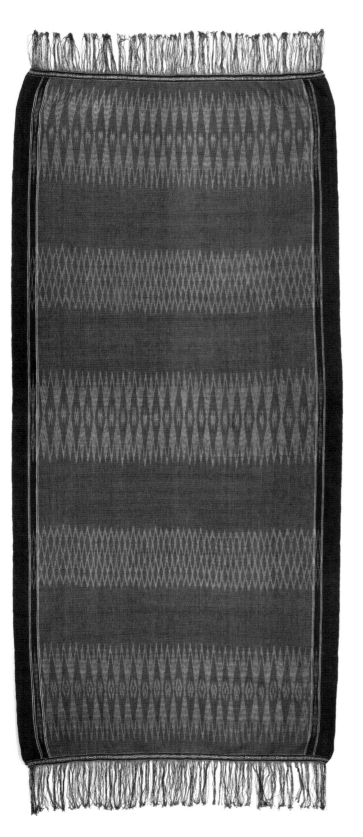

Fig. 10.4 Batak ceremonial shoulder cloth (*sibolang*) cotton with warp-ikat patterning, warp stripes, and weft twining, 105 cm x 237 cm. Private collection.

Fig. 10.5 Karo Batak weaver, North Sumatra, Indonesia, c. 1925 (photo: Leiden University Libraries, KITLV 5533).

tradition of weaving with *abaca* is still continuing until today (fig. 10.1 and facing page).[6] The ground weave produced on the body-tensioned loom is mostly plain weave, with rare exceptions of some twill weaves.[7] The two sheds needed to make the plain weave are opened by a bamboo bar for the natural shed and by a string heddle for the counter shed. The cloth width—limited to the outstretch of the weavers' arms—can surprisingly be very wide, as much as 1 m as in the weaving of the Bataks of North Sumatra (figs. 10.4 and 10.5). In order to maintain an even width of the woven cloth, weavers insert the pointed ends of a bamboo stick called a temple near the selvedges as the weaving progresses.[8] In some body-tensioned looms, a warp-spacer device in the form of a notched rod may be placed near the warp beam.

The weave produced on this simple body-tensioned loom is inherently warp-faced. Thus, the loom is primarily used for warp patterning,

83

Fig. 10.6 *Mawa* from the Toraja region of Central Sulawesi, Indonesia, 19th–20th centuries. Cotton plain weave, painted. The form is highly unusual because it consists of two seamless circular cloths that interlock like a chain. Here only one side of the cloths is shown, 40.6 cm × 40.6 cm. (photo: Cooper Hewitt Smithsonian Design Museum, New York, 1984-20-1).

weaving traditions in Sulawesi where the unwoven warps are completed by using a needle to weave in the weft. Examples are the *pinatikan*, a warp-float patterned textile,[11] and a ritual textile called *mawa* (fig. 10.6). There are also temporary practices where tubular textiles may not be cut. For example, the bridewealth sarong from Lamalera (*kewatek*)—a symbolic ceremonial item not intended to be worn—is constructed of three pieces of uncut tubular cloths that are sewn together along the selvedges to make a longer tube.[12] The Balinese *geringsing wayang* double ikat is another example where the circular form is significant to the textile's function: only the uncut and unused *geringsing* can be presented as offerings to the divines.[13]

including warp stripes, warp ikat, supplementary warp, and warp floats. Skillful weavers, however, are able to overcome the loom's natural limitations and create a variety of different decorative techniques. Weavers in the Tenganan Pegringsingan village in Bali, for example, weave double ikat on this loom, even though this technique requires a balanced plain weave in order to make the ikat pattern on both the warp and the weft visible in the finished textile (cat. 30). In Timor and Borneo, weavers create a weft-patterning effect (called *buna* and *sungkit* respectively) by wrapping together several warps with discontinuous supplementary wefts.[9]

When the cloth is taken off the loom, the finished product is a tubular cloth with a narrow section of unwoven warp. In most cases, these open warps would then be cut.[10] There are, however, early

10.2 Reeds

The introduction of reed opened up new possibilities for creating different types of weaves and decorative effects. In the body-tensioned loom, reed functions primarily as a warp spacer. This allows the weaving of fine fibers such as silk, weft-oriented patternings such as weft ikat and supplementary weft, and balanced weaves such as plaid.[14] Reed on the body-tensioned loom is also used together with the sword as a beater but never on its own, as in the frame loom.

In Indonesia, the reed is associated with weaving technology that appeared during the Hindu-Buddhist Period (5th–15th centuries). Certainly by the 10th century, Old Javanese and Old Balinese inscriptions had already listed reeds (*suri*) as one of the items sold in the market. Inscriptions from the same period also mention a type of specialized loom, *pacadaran*; the weavers of this loom (*acadar*) belonged to a professional group of crafts people.[15]

The oldest visual evidence of the use of reed is found on a weft ikat, which belongs to a group of textiles with a radiocarbon date of the mid-15th century (fig. 10.7).[16] This weft-ikat group may represent a transition in the ikat traditions from warp to weft, facilitated by the reed technology.

10. Insular Southeast Asian Looms

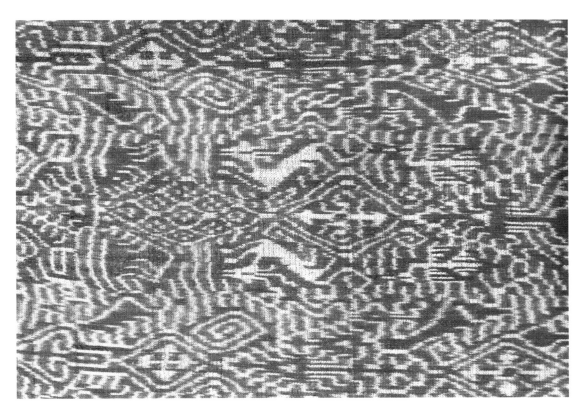

Fig. 10.7 One of the oldest weft ikat textiles found in Indonesia. Plain weave with silk warp and cotton weft. Former collection of the Nusantara Museum, Delft (photo: Sandra Sardjono). Above: detail of the intricate ikat pattern showing deer and birds. Below: 1 cm square area of the cloth, showing paired warps with gaps between that indicate the use of a reed.

Fig. 10.8 Ni Gede Diari weaving cotton *cepuk*, Nusa Penida, Bali, Indonesia, 2018 (photo: Sandra Sardjono).

10.3 Half-Frame Looms with Reed and Flat Warp

The body-tensioned loom with a reed usually has a non-circular warp or *flat warp*. Here the warp beam may be round or may take the shape of a plank of wood board onto which one end of the warp length is wound. When the warp beam is lodged into a slot at the back of a low frame, while the cloth beam is still tied to the weaver's waist, the loom may also be called the half-frame loom. Such loom type is found in Sumatra, Java, Bali, Sumbawa, and South Sulawesi (fig. 10.8, cat. 33). The flat warp system allows practically unlimited length of warp to be prepared at once, which saves a lot of time. In the case of weaving the Nusa Penida *cepuk* shown in fig. 10.8, the weaver has prepared a warp length enough to make four to five cepuk cloths.

Although reed is primarily associated with silk and weft patterning, it is also used for weaving cotton and warp patterning, as in the cotton warp ikat ceremonial wall hangings (*kasang*) from West Java. The *kasang* loom signals the merging of the older technology of cotton and warp ikat with the newer technology of silk and weft ikat. Other examples of blended traditions that use the body-tensioned loom with reed and flat warp are the silk warp ikat from Aceh in North Sumatra and the cotton weft ikat from Bali and Lombok.

Fig. 10.9 Ni Gede Diari pulling loops of warps into a reed for weaving *cepuk*, Nusa Penida, 2018 (photo: Ngurah Hendrawan).

10.4 Body-Tensioned Looms with Reed and "False-Circular" Warp

The use of a reed does not naturally lend itself to a circular warp because the yarns have to be inserted through the comb of the reed (fig. 10.9). But there are body-tensioned looms equipped with a reed in eastern Indonesia and southern Philippines that appear to retain a circular warp. Fig. 10.10 shows two weavers in Nggela village in Ende, Flores, warping this loom for a man's sarong.

The warp was drawn from a ball of yarn as a loop, meaning each draw of the yarn creates a double thread. The younger weaver slides one end of the loop onto the warp beam and makes the heddles; the older weaver threads the other end of the warp loop through the opening of the reed and then slides the end onto a rod. The next time she takes another warp loop, she passes it under the reed and slides it onto the same rod. In other words, this rod connects the ends that pass through the reeds and those that do not, making a seemingly circular warp during weaving. When the "connecting rod" is withdrawn after the weaving is completed, however, the tubular form of the woven cloth opens

Fig. 10.10 Left: two weavers in Flores, Indonesia, preparing a false-circular warp for a body-tensioned loom with a reed. The younger woman is holding a white string that she uses for making the string heddles. The older woman is preparing the reed. She inserts loops of warps only on the top layer of the warps, 1980 (photo: Kajitani Nobuko).

Right: detail of the loom, now in the collection of the Fowler Museum at UCLA, 2015-9-10. Seen on the photograph are the cloth beam, reed, sword, heddle rods for the counter shed and supplementary-wefts, and rod for the natural shed (photo: Sandra Sardjono).[17]

up into a flat textile.[18] This warping method is illustrated by Jasper and Pirngadie, in their seminal study, *De Weefkunst* (fig. 10.11).[19]

The loom with a false-circular warp is found in the eastern part of Insular Southeast Asia, encompassing the islands of Sangihe, Salayar, Talaud, Sula, North Maluku, the western part of Flores, and Sulawesi (especially the northern and southeastern parts of the Peninsula along the Gulf of Bone) as well as the eastern Mindanao (used by the Mandaya and Manobo people). Yoshimoto Shinobu has suggested that that the false-circular warp technique is particular to some parts of Sulawesi and nearby areas that had close historical contact with the island.[20]

10.5 Frame Looms with Reed and Flat Warp

A frame loom, equipped with a reed and a flat warp, is suitable for the making of textiles with weft-oriented patterning and for weaving silk. Frame loom weavers normally sit on a bench, and they operate the sheds using foot treadles instead of lifting heddles by hand.

In Indonesia, frame looms were already used in the Minangkabau region in West Sumatra by the 19th century (fig. 10.12). One of its earliest rendering is published by P. J. Veth in 1882.[21] The origin of the Minangkabau loom is uncertain, but it bears close similarities with those used in Malaysia. This is not unexpected due to the close historical connection between the northern Sumatra and the Malay courts, as well as their shared love for textiles with gold thread supplementary-weft weaving (*songket*), which are produced on these frame looms. The Malay frame loom itself may have been derived from the Cambodian Khmer frame loom.[22]

There is also a smaller type of frame loom for weaving narrow bands, which has been recorded in use in Bali, Lombok, and Java. The weaver of this loom may sit or stand by the side of the loom. Characteristics of this loom are the pattern sheds that are weighted and the heddle-treadle mechanism that is movable. The distribution of this loom appears to have been much wider in the past, including parts of the Asian mainland.[23]

In the Philippines, weavings on frame looms were becoming more common after World War I. A unique feature of this loom is the arrangement of the warp, which goes upward from the back beam and winds around a large polygonal wooden bar (fig. 10.13).

Fig. 10.11 Warping of loom with false-circular warp in Sulawesi. The second bar from the left is the "connecting rod" (after Jasper and Pirngadie, 1912, fig. 80).

Fig. 10.12 Weavers in Kota Gadang, West Sumatra, c. 1895 (photo: Wereldculturen, the Netherlands, TM-60003539).

Fig. 10.13 Tboli weaver on a frame loom, normally used by the Maguindanao weavers. Barangay of Sinolon, 2016 (photo: Craig Diamond).

10.6 Temporary and Permanent Pattern-Shed Saving Devices

To create a woven pattern—whether it is a float, wrapping, supplementary, or complementary technique—requires the lifting of specific warps for the pattern sheds in addition to those for the ground sheds. Weavers could either pick the warps by hand or by using a device that stores the pattern shed lifts so they can be easily repeated. There are two main systems for pattern-shed saving devices in Insular Southeast Asia: a warp-rod system and a heddle-rod system. Each system can be used on its own or in combination.

The warp-rod system is a temporary pattern-shed saving device. It consists of a series of rods that interlace with the warps to mark the warp lifts for the pattern sheds. First, a weaver has to interlace these rods into the warps. Then she lifts the rod closest to her, inserts the sword into the pattern shed to enlarge the opening, and throws the shuttle across the shed. After a rod is used, it must be removed before the weaver can use the next one. When all the rods have been used, the weaver has to re-insert the rods again for the next pattern sequence. On a body-tensioned loom, the rod system is popular for making supplementary-warp and supplementary-weft patternings. A notable example is the supplementary-warp patterning from Sumba called *pahikung* (fig. 10.14).

Specific to Sumba, the interlacing sequence of the pattern rods is also recorded on a separate pattern guide (*pahudu*), which consists of a set of threads interlaced with sticks (fig. 10.15).[24] The pattern guide allows weavers to recreate the same exact pattern on a different loom or on a different area of the weaving.[25]

Unlike the warp-rod system, the heddle-rod system is permanently fixed on the loom. The heddles, which are made of strings, loop around specific warps. Each heddle-rod represents the warp lifts of one pattern shed; thus, the number of rods increases as the pattern grows in complexity. With this permanent pattern-shed saving device, a weaver is able to use the heddle-rods repeatedly and in any order because each one is independent from the others.

In the body-tensioned loom and the half-frame loom weavers can operate both systems by themselves because they can reach the rods and heddles by leaning forward and manipulating the tension of the warp. This is not the case with the frame loom. As shown in the Minangkabau loom (fig. 10.12), a weaver would need an assistant to reach and operate the pattern-shed device, which extends from behind the ground-weave heddles towards the back of the warps.[26]

Some textiles are woven using both the warp-rod and heddle-rod systems. An example is the warp-float patterned *pinatikan* from Minahasa, Sulawesi. The *pinatikan* loom is a body-tensioned loom with circular warp. It has a set of pattern heddle-rods, which is not used directly; first, the weavers "transfer" the sheds of the heddle rods onto temporary warp-rods. These warp-rods are then used to open the sheds for the weft insertions. Rita Bolland suggests that the two-step process may have to do with creating a particular rhythm that assists the mechanical process of weaving and prevents the weavers from making mistakes.[27] This double mechanism can be seen in the supplementary-weft patterning as well, notably in the *songket* weaving tradition, which uses the half-frame loom with reed and flat warp (figs. 10.16 and 11.1). In the case of *songket* weaving, the heddle-rods can be numerous, with hundreds of rods. The use of temporary pattern rods allows the weaver to selectively pick out and concentrate on the groups of pattern sheds that are needed at the moment while setting aside the rest (cat. 33).

Fig. 10.14 A weaver making *pahikung* in Pao village, East Sumba, Indonesia, 2017 (photo: Serena Lee).

Fig. 10.15 A pattern guide from Pao village, Sumba, 2018 (photo: Sandra Sardjono).

Notes

1 Buckley 2012; Buckley 2017.

2 For the Austronesians, the cosmos and society as its mirror were organized based on a system of duality, in which at the most fundamental level lies the duality of male and female. Men hunt and work with male specific-items such as metals, while women create female items such as cloths.

3 Robyn Maxwell's study (Maxwell 2003) is one the most serious endeavors so far to show how the diverse Southeast Asian textile traditions relate to one another.

4 This introductory essay focuses more on looms from Indonesia than any other region, which reflects the current state of research of the textile field itself.

5 The more widespread external-brace system allows the warp beam to be extended further beyond the limited stretch of the weaver's legs. In addition, the system enables the warp beam to be placed higher than the cloth beam; this elevated position creates a downward slant towards the weaver, which takes advantage of the force of gravity when she beats in the wefts with the sword.

6 For more discussion on the *abaca* warp-ikat tradition in the Philippines, see Hamilton 1988 and Pastor-Roces 1991. I am thankful to Dr. Michael Gonzalez for his insight on Phillipine looms.

7 The Tropenmuseum, Amsterdam, owns an early 20th century loom from Borneo with twill weave in diamond patterns. See discussion on the significance of twill weavings in Southeast Asia in Van Hout (forthcoming).

8 A temple is sometimes inserted on the top of the fabric and sometimes underneath.

9 See discussions on *sungkit* (Gavin 1996, 70–75) and on *buna* (Barnes 2014).

10 There are several ways tubular cloths may be joined to make different articles of clothing (Hamilton ed. 1994, 76).

11 Bolland 1977.

12 Barnes 1989.

13 On the topic of circular and uncut geringsing as offerings, see Hauser-Schäublin, Nabholz-Kartaschoff, and Ramseyer 1991,124.

14 As in the example of Balinese double ikat, however, not all balanced weaves require a reed.

15 It is unclear how the reed was introduced into Java. Christie suggested that it may have to do with the contact with the Cholas in South India. In this case, professional Muslim weavers using special loom with reed (achchutari) were already mentioned in Tamil inscription dated from the late first millennium CE. The fact that these professional weavers are Muslim also suggested to her that looms with reeds originated from Persia (Christie 2000, 22; Christie 1993, 189).

16 One published piece in this group was collected from the Komering river area in South Sumatra. (Barnes ed. 2010, 48–49). Several other published pieces are in the collection of the Department of Indo-Pacific Art, Yale Art Gallery (Barnes 2018).

17 I am grateful to Joanna Barrkman, Senior Curator of Southeast Asian and Pacific Arts, Fowler Museum at UCLA, for organizing the viewing of this loom in February 2018.

18 The process is also described in Hamilton 1994, 70–71.

19 Jasper and Pirngadie 1912, 101.

20 Yoshimoto 1990.

21 Veth 1881, plate CXV. I am grateful to Bernhard Bart for pointing out this reference. This image is also published in Summerfield Anne, Summerfield John, and Abdullah 1999, 211.

22 Bernhard Bart (forthcoming) proposes that Khmer-loom influence in Malaysia arrived with the Cham migrants.

23 This loom is depicted in Gittinger 1979, fig. 4.

24 The methods of "storing" woven patterns in different areas is an interesting subject worth exploring further. The author was told by a Sumba weaver in early 2019 that *pahudu* is normally made by experienced weavers for use by less-skilled weavers. Hamilton also describes that experienced weavers and less-skilled weavers in Flores have different approaches in dealing with supplementary-weft weavings (Hamilton 1994, 73).

25 If the pattern is symmetrical, only half of the warp lifts need be picked out by hand. The weaver would insert two pattern rods at the same time, one would be used for the top and the other the bottom, and she would work outwards from the middle.

26 Weavers' dependency on an assistant is eliminated in frame looms with vertical pattern heddles, such as those found in Mainland Southeast Asia and Southwest China (sections 6.6 and 9.4, cat. 26)

27 Bolland 1977, 4.

Fig. 10.16 Facing page: silk ceremonial cloth (*limar*) from Palembang area, South Sumatra. Silk weft ikat and supplementary-weft patterning with gold thread (*songket*). Private collection.

11. Voice of the Weaver: Palembang and Muang Kham

Zainal Arifin, Director of Zainal Songket, is from the 3rd generation of weavers from Palembang, Indonesia (fig. 11.1). He is one of eight siblings, all of whom took up the weaving profession after their parents, Kgs. H. Husin Rahman and Nyayu Hj. Fatimah Cek Ipa. Arifin learned to weave from the age of eight. He specializes in the famous Palembang *songket*, silk with gold and silver supplementary-weft patterns..

I am a specialist in songket *weaving. I have been weaving for 35 years. I love culture, and* songket *is my cultural heritage. I am the third generation of weavers. The tradition has been passed on from my grandmother, mother, and now me. But my children are not interested. I am the last generation of weavers in my family.*

Fig. 11.1 Zainal Arifin demonstrating weaving (photo: Sandra Sardjono).

Bouakham Phengmixay (Inthavong), comes from Ban Natoum, Kham district, Xiengkhouang province, northern Laos (fig. 11.2). Now in her 40s, she is descended from a long line of female weavers. Her grandmother is originally from Houaphan, Hiem village from the Thai Chao ethnic group. Phengmixay was taught basic weaving using cotton by her mother at age seven. By age twelve she was skilled enough to pick out a design, make the heddle, prepare the warp and dye natural colors. At the age of sixteen, she was able to weave silk skirts. Her grandmother encouraged her to work with weaving and taught her to sell her textiles. In 1997, Phengmixay started to work as a skilled weaver at Lao Textiles, Vientiane, Laos.

Weaving in this loom and this technique is special because it's from my ancestors, and from my grandmother to my mother, and it's also a custom that all women in my culture have to weave. So then it comes to me and I like to weave because it is also for the Lao women to weave. And I like to hear the voice of the weaving, like when I hit the comb and when I press the bamboo down, and everything about it... that's what I love about the weaving.

Fig. 11.2 Bouakham Phengmixay with Carol Cassidy during weaving demonstration (photo: Sandra Sardjono).

South Asia

12. Indian Looms
    Hemang Agrawal

13. Voice of the Weaver: Varanasi
    Naseem Ahmad

12. Indian Looms

Hemang Agrawal

Stories of life and loom have intertwined for centuries in India. The following lines by the 15th century Indian philosopher Kabir who was also a weaver (fig. 12.1), give insight into this relationship:

How delicate and pure is this fabric the Lord has woven?
What is this warp? What weft is this? Of which fiber was this fabric made?
The breath taken is this warp, the breath released is the weft; the veins that stretch, the nerves that bind, their flow is this very fabric.
An eight-petaled lotus is the Lord's spinning wheel; it spun the five elements, and all three virtues, from which this fabric was woven.
It took the Lord ten months; it took the finest skill, and infinite patience; it took endless labor to beat-in this weft.
Worn by celestials, worn by seers and mortals alike, this fabric was defiled by one and all; your devotee Kabir wore it scrupulously and discards it just as it was: forever unsoiled, forever pure.[1]

Kabir lived in the famous Indian city of Benares (Varanasi), which is considered to be the textile capital of India. In this poem, he likens God to a weaver who created the bodies of humankind on his loom. Kabir thus identifies the act of weaving with the act of God's creation itself. The poem is an indication of sanctity with which hand weaving is regarded in India. Up to the present day, India remains a bastion of hand weaving, perhaps the last place where handlooms continue to operate on an industrial scale.

Fig. 12.1 Kabir with disciple, weaving cloth. Painting from c. 1825, Collection of the Maharaja Sawai Man Singh II Museum, Jaipur.

12.1 Handloom Landscape in India

According to the Government of India's handloom census in 2010, the total number of handlooms operational in India was more than 2 million. In 2014, there were 6 mega-clusters having more than 25,000 handlooms; 20 clusters with 5,000 handlooms and 600 clusters having 300–500

Facing page: Paithani saree, with tapestry and supplementary-weft patterning, India, 20th century.

Fig. 12.2 The logo of the "India Handloom" brand, a national seal for handloom products from India.

The eastern and the adjoining coastal regions produce a gamut of textiles. The hand-inserted supplementary weft (inlay) cotton *jamdani* of Bengal, the *bandha* single-ikat of Orissa and the pictorial silk *baluchari* of Murshidabad near the Bangladesh border, woven using a *jaala* or a Jacquard attachment[2], are all painstakingly produced over months of work, and are a testament to the high level of skill of the weavers of this region. This region also produces a variety of fabrics woven with wild silk yarns on fly-shuttle looms.

Over the last few decades, the Deccan and southern states of India have emerged as not only the biggest producer but also the biggest consumption center of the staple product of Indian handlooms: the saree. Besides the well-known *Kanjeevaram* saree, which is woven with a *jaala*-like *adai* patterning device, the region also produces several other famous sarees including the *Pochampally* silk ikats, the *Mangalgiri* cottons, the silk-cotton *Gadwals*, *Chettinad* cottons, *Dharmavaram* silks, and the *Kerala Kasavu* cottons. Most of these sarees are now woven on frame looms with a dobby or a Jacquard attachment for the borders and the end-pieces, under the aegis of the large, co-operative weaving societies spread over southern India. Although the metallic yarn (*zari*) is an integral part of Varanasi and several other weaving traditions of India, it is in southern Indian sarees where one sees its most widespread use.

The western and central states, well known for their resist-dyed textiles, are also equally famous for woven textiles. The *paithani* sarees from Maharashtra with their double interlocking tapestry weave, the double-ikat *patola* sarees from Gujarat with their unique reed-less loom, which is aligned at a shallow angle to the horizontal, and the *Chanderi* sarees from Madhya Pradesh now woven on fly-shuttle frame-looms equipped with a *jaala* or Jacquard, are among the most striking textiles from this region. Other notable handwoven textiles from the area are the *yeola* sarees, *tangalia* shawls, *asavali* brocade sarees and *khan* blouse textiles.

The cities of Ahmedabad and Surat in Gujarat had a flourishing silk-weaving tradition during

handlooms each. These numbers underline an important point: while hand weaving is now largely confined to artisanal studios elsewhere in the world, in India handlooms produce not just artisanal but also commodity textiles of a reasonably high quality, for traditional ceremonial usage, casual wear, and home furnishings (fig. 12.2).

Paralleling the diversity of dialects, which number around 1,600, the textile traditions in India have distinguishing features every hundred or so kilometers. The regional differences are often found in loom typology and almost always in the fabric types and end products.

The seven sister-states forming Northeast India are the cradle of body-tensioned looms, also known locally as backstrap loin looms. The weavers, who are predominantly women homemakers, produce a variety of narrow-width cotton textiles for daily as well as ceremonial usage. This region also has an abundance of throw-shuttle frame looms, which produce *mekhala*, *moirangphee* and other varieties of wraps (*chador*) with a plain weave ground and basic supplementary weft patterning, in both cotton and silk.

the Mughal Period, which is believed to be the origin of the complex pattern-weaving traditions of Varanasi. The technology of these looms has links with Persian and Central Asian traditions, particularly in the use of horizontal cross-cords suspended over the warp (*pagia*) for controlling warp-lifts, to which patterning systems can be attached.

The most notable of the wool-weaving traditions in India belong to the upland regions of North India, particularly Kashmir and Himachal, where pit looms with two or four treadles predominate. With their double-interlock tapestry weave using long wooden spools (*kani*) in place of bobbins, worked in a fine twill foundation, the *kani pashmina* shawls of Kashmir have been among India's best-known textiles for the past three centuries. The Himachal region, on the other hand, produces a variety of heavier shawls, wraps and yardages in coarser wool with simple tapestry patterning.

The northern Gangetic plains of India are home to the most delicate as well as intricate woven textiles. The *awadh jamdani*, with its subtle white-on-white cotton inlay, comes from the Gangetic plains of the state of Uttar Pradesh. There is also an abundance of vertical and horizontal ground looms for weaving coarse floor-covering textiles in this region. However, the region is most famous for the *Benares* brocades from Varanasi, which represent the zenith of modern Indian patterned weaving. Varanasi continues to use a variety of patterning techniques such as the hand-manipulated *jamdani* inlay, paper pattern-assisted *uchinna*, the multi-shaft *gatthua* with its complex twill treadling, the "modern" Jacquard mechanism, as well as the ancient cross-cord *pagia* pattern harness with its *jaala-naqsha* mechanism. Products include sarees, wraps and yardage.

12.2 Loom Typology in India

The majority of Indian handlooms are simple, two or four-shaft frame looms. These are, however, very diverse in terms of typology, operation, fabric design as well as the end product. At the other end of the spectrum lie the complex looms that produce India's famed patterned silks. These incorporate the Jacquard mechanism as well as several older drawloom mechanisms.

A feature of the Indian handloom landscape is how the weaving techniques developed over centuries continue alongside more recently developed technologies, deploying tools and practices that have largely remained unchanged over time. A number of these techniques are probably indigenous while others have been acquired from elsewhere. The growth and development of weaving technologies is not linear and hence the classifications presented here, from simple to complex, does not necessarily reflect the chronological development.

12.2.1 Body-Tensioned Loom

The loin or backstrap loom (fig. 12.3) is a reed-less loom that relies upon body tensioning from the weaver. These types of looms are prevalent in Northeast India and are related to their better-known Southeast Asian counterparts. Weavers, who are all women, use these looms to produce narrow-width textiles for daily or ceremonial use, often with supplementary weft patterning.

Fig. 12.3 Weaving on a body-tensioned loom in Northeast India.

12.2.2 Vertical Loom

The vertical reed-less loom is used to a limited extent for making coarse cotton durries and floor-coverings (fig. 12.4). Geometric patterns are woven in the textiles using dovetailed and slit tapestry weaves. The *navalgund jamkhanas* floor-coverings from Karnataka are a prime example of the products from this type of loom.

12.2.3 Horizontal Ground Loom

The horizontal counterpart of the *jamkhana* loom is used more widely in India. Across northern India horizontal ground looms (fig. 12.5) produce commercial floor-coverings in varied sizes and large quantities. In both vertical and horizontal looms, the patterning is done using dovetailed and slit-tapestry weaving. The beating-in of wefts is done using a metal fork, which is locally called a *punja*. Hence these floor coverings are known as *punja-durries*.

12.2.4 Patola Loom

Somewhere between the body-tensioned and the horizontal ground loom sits the frameless *patola* loom (fig. 12.6) from the town of Patan, Gujarat. It is again a reed-less loom, with a differentiating characteristic of being tilted at a slight angle to the horizontal. The extremely time-intensive double ikat *patola* saree is woven in plain weave. The loom requires two weavers to operate it, and in absence of a reed, the beating-in of wefts is done using a wooden sword.

Above: fig. 12.4 Carpet weaving on a vertical loom.

Middle: fig. 12.5 Carpet weaving on a horizontal ground loom (photo: Roopraj Durry Udhyog).

Below: fig. 12.6 *Patola* loom in Patan, Gujarat.

12. Indian Looms

12.2.5 Pit Loom

The simplest and the most common of Indian looms is the pit-loom with treadles (fig. 12.7, cat. 34), mainly used for weaving plain weave cloth made from handspun cotton (*khadi*) in Bengal and southern India. Throw-shuttle pit looms are suited to handling finer grades of cotton yarn in the weft, and it can be presumed that this loom type is indigenous. The simplest pit looms use throw-shuttles but there are also looms with fly-shuttles, a European invention that has been widely adopted since the 19th century.

12.2.6 Simple Frame Loom

With the emergence of organized cottage industries, especially in southern India, one often comes across the counter-balanced, fly-shuttle frame looms for weaving plain weave cottons. As with the backstrap loom, women are the main users (fig. 12.8), and today women constitute more than 75% of the total Indian weaving workforce.

Fig. 12.7 Weaver operating a pit loom in Rajasthan.

Several of the ornamental but non-patterned cotton and silk saree weaving traditions of India, specially those of resist yarn-dyed sarees in the southern and eastern-coastal regions use variations of simple pit or frame looms.

Fig. 12.8 Weaving on a frame loom.

Fig. 12.9 Weavers in Bengal weaving a *jamdani* saree.

Fig. 12.10 Manual selection of warp yarns for patterning in a Pashmina shawl.

12.2.7 Pit or Frame Loom with Hand-Inserted Jamdani *Brocading*

While the term *jamdani* is used to identify the fine white-on-white "inlay" patterned textiles from Dhaka and Tanda, the term can also be used for several other Indian supplementary-weft textiles such as *tangalia*, *chanderi* and *venkatgiri* (fig. 12.9), and this was how almost all patterning was done in India before the advent of *jaala* or Jacquard devices.

In the *jamdani* technique, the lifting of the warp ends for weft insertion is done either from memory, as practiced in Dhaka and Tanda, or by using a paper pattern under the warp, termed an *uchinna*, as practiced in Varanasi and elsewhere. The supplementary weft travels in the same shed as the body weft and is called *dampach*. Even the Paithani tapestry weaving of West India, although a complimentary weft-patterning technique, is hand-manipulated in a similar way to *jamdani*. Commercial production of *jamdani* textiles is mostly done on fly-shuttle looms. In this case the supplementary weft travels the entire width of the cloth (*fekuan*) and the floats at the back are subsequently sheared (*katarwaan*).

12.2.8 Frame Loom for Pashmina Shawls *(*Kani*)*

Kani pashmina shawls from the Kashmir region have been among India's best-known textiles for the past three centuries, as well as major exports to Western markets. The shawls are worked in a fine twill foundation with double-interlock tapestry weave for patterning, using long wooden spools called *kani* in place of bobbins (fig. 12.10). This type of weaving also deploys a numbered lifting-sheet as a guide, which in former times was recited as a poem by the master weavers to guide the assistants in lifting warps.

12.2.9 Loom with Jacquard Dobby for Bordered Fabrics

These looms find widespread use in South Indian co-operative societies, where the dobby attachment is deployed to weave borders, usually with extra warp of metallic yarn (fig. 12.11).

Fig. 12.11 Loom equipped with a dobby patterning system.

Most saree-weaving traditions of South India like *Pochamplally*, *Mangalgiri*, *Kerala* kasavu require minimal brocading (decorating with supplementary weft) contained only within the borders. The use of the dobby attachment greatly increases weaving speed.

12.2.10 Pit Loom with Simple Jaala

Cross-cord pattern harness systems (referred to as *pagia* in some regions) are used in several parts of India. The Aurangabad *himru*, Kanchipuram *adaai*, the Varanasi *gethua* and the Chanderi *jaala* all deploy lifting mechanisms, which can be viewed as simpler versions of the Varanasi *jaala*. These simpler systems are used for less intricate motifs and have several variations. In Chanderi, Madhya Pradesh, a small *pagia* is tied in front of the body shafts, rather than behind, which makes for easier lifting by the weaver himself, without requiring a drawperson (fig. 12.12). The cotton or silk-

Fig. 12.12 Weaver manipulating heddles in a pit loom with *jaala* attachment.

cotton sarees thus produced usually have stepped geometrical patterning using "inlaid" tapestry and supplementary weft brocading.

12.2.11 Loom with Hand-Operated Jacquard

The hand-operated Jacquards are the most widely used patterning mechanisms in India today (fig. 12.13). In Varanasi, handlooms with Jacquards of up to 400 hooks are a common sight and are usually operated by a single weaver. For custom production, Jacquards with up to 1,000 hooks are still in use.

The entire process from paper-pattern designing, to graphing, to punching and lacing cards together is done manually, so despite the Jacquard mechanism's associations with "automation," this type of weaving remains firmly in the traditional, handmade sphere. The Jacquard attachment facilitates single-weaver operation; this advantage is one of the main reasons why Varanasi became a patterned-textile weaving center on an industrial scale in the 20th century. The textiles that are produced range from highly detailed satin silk and plain weave brocaded sarees to ornate yardages for garments and luxurious home furnishing.

12.2.12 Pit Loom with Complex Naqsha-Jaala

The classical Indian drawloom with a cross-cord pattern harness, as used in Varanasi is called the *naqsha-jaala* loom (figs. 12.14–12.16, cat. 36). The term *jaala* translates to "a network of threads" which act as the pattern storage as well as the pattern-harness of the drawloom. The *jaala* tradition in India traces its origins to a 14th century master, Khwaja Bahauddin Naqshband of Bukhara, Central Asia. By the 16th century, one comes across references to textiles that are similar to the Iranian textiles produced on drawloom being woven at the Mughal workshops or *karkhanas* in Gujarat, western India. It is from these Mughal *karkhanas* of Gujarat that the complex *jaala* drawloom mechanism reached Varanasi in the early 18th century.

The key component of this drawloom mechanism

Fig. 12.13 Loom with a hand-operated Jacquard mechanism.

12. Indian Looms

Fig. 12.14 Weaving on a *jaala* drawloom at the workshop of Rahul Jain, Varanasi, India (photo: Carol Ventura).

Fig. 12.15 *Naqshband* (pattern master) Naseem Ahmad preparing a new design for a *jaala* loom.

Fig. 12.16 *Jaala* drawloom in operation.

is the *naqsha*, a detachable set of vertical drawcords and horizontal pattern lashes, which acts as a template for pattern lifting. A hand-drawn design is transferred onto a graph, which is then used as a guide by the *naqshband* (*naqsha* master) to make the *naqsha*. This is prepared on a frame called *machaan* and then installed at the upper level of the drawloom.

The Varanasi *jaala* drawloom (fig. 12.16) differs from other traditional Indian looms by the presence of a warp beam and the absence of a counter-balance mechanism, hence there are separate sets of treadles to control the lifting and the depression shafts.

In the *jaala* loom set-up, the harness structure consists of jack-type wooden shafts while the pattern harness comprises of horizontal cross-cords (*pagia*), which hold groups of warp ends with string leashes called *naaka*. The vertical drawcords of the naqsha are knotted onto these horizontal cross-cords.

To operate the *jaala* drawloom, the drawperson pulls a single pattern lash of the *naqsha* to separate the drawcords for that particular pattern pick and pulls these draw-cords by inserting and twisting a wooden fork that is locally called a mantha. These pulled draw-cords lift their corresponding cross-cords and thereby the string leashes and the warp-ends connected to them, thus creating a patterning shed to pass a pattern weft through. The weaver then inserts an angled-hook called *akda* under the cross-cords to equalize the shed height across the loom width and also keep the shed open once the drawperson releases the lift. In contrast to the noisy clitter-clatter of Jacquards, *jaala* weaving is virtually silent.

Famous Indian gold and silver-brocaded textiles include sashes, sarees, and wrappers, alongside fabric lengths for stitched garments and furnishings, the heyday of which was in the 16th to the 19th centuries. All owed their magnificence to the exceptional skills of the Indian *jaala* drawloom weavers operating in the Mughal *karkhanas* of Western and Northern India, Varanasi and a few other weaving centers.

Several types of Indian brocaded textiles (fig. 12.17) were woven using complex double-warp weaving techniques such as lampas, taqueté and samit. The second set of warps was used to bind the supplementary weft floats, resulting in a plain surface at the back. The complex pile-warp velvet technique is the most complex form of weaving which deploys the *jaala* mechanism.

Although the usage of complex *jaala* mechanism in Varanasi has diminished over the last few decades, it continues to be used owing to its adaptability and versatility. *Jaala* is sporadically used in conjunction with a Jacquard attachment to weave a saree, in which case the Jacquard is used for the border and the body and the *jaala* is used for a *konia*, which is a diagonal paisley motif at the end-piece of the saree.

All the major classes of fabrics in Varanasi incorporate metallic yarns of gold and silver, and for that reason Benares has traditionally been the producer of a very high quality of silver yarn with 99% purity.

India's handwoven and handcrafted textile traditions and their associated weavers form perhaps the country's most important cultural demographic. They are a true manifestation of India's "unity in diversity and diversity in unity." It can be said that the future of handlooms around the world is inextricably linked to their future in India.

Fig. 12.17 Silk textile woven on a *jaala* loom.

Notes

1 *jheeni-jheeni beeni chadariya*

kaahe ka tana, kaahe ki bharni, kaun taar se bini chadariya

ingla-pingla taana-bharni susman taar se bini chadariya

asht-kamal dal charkha dole, paanch tatva, gun teeni chadariya

sai ko bunat maas das laage thok thok ke bini chadariya

ye chadar sur, nar, muni odhi, odh-odh kini maili chadariya; das Kabir yatan se odhi, jyon ki tyon dhar deeni chadariya.

2 Such looms are conventionally called "Jacquard looms," though in fact the technology used embodies several advances versus the original Jacquard loom (section 24).

13. Voice of the Weaver: Varanasi

Naseem Ahmad comes from Varanasi, India. Just like his late great grandfather Ali Hasan and his grandfather Jafar Ali, he continues the tradition of Banarasi *saree* weaving on *jaala* looms. Today he is in his mid-forties, and he is the last remaining custodian of this intangible heritage (figs. 13.1 and 13.2).

My weaving needs some extraordinary skills, just like how to make… jaala [weavings] on the loom. This is very tough; we have to take care of the directions of the design—right or left—to be woven… And the number of threads must be equal according to our layout of the design. It is very important for us to manage all the necessary arrangements on the loom set-up… That's why you called me a magician, but in fact I'm a hard worker!

Fig. 13.1 From left: Naseem Ahmad, Tauseef Ahmad Ansari, and Hemang Agrawal with a model *jaala* loom (photo: Elena Phipps).

13. Voice of the Weaver: Varanasi

Fig. 13.2 Varanasi weavers and museum staff preparing the warp for the *jaala* loom (photo: Elena Phipps).

Central and Southwest Asia

14. Central and West Asian Looms
    Gillian Vogelsang-Eastwood

15. Traditional Weaving of Uzbekistan
    Binafsha Nodir

16. Voice of the Weaver: Margilan and Ardakan
    Rasuljon Mirzaahmedov, Rastjouy Ardakani Mirzamohammad

14. Central and West Asian Looms

Gillian Vogelsang-Eastwood

The terrains in Central and West Asia include mountainous areas, steppes, deserts and semi-deserts. Such regions are ideal for raising flocks of sheep and goats, as well as animals such as camels and horses for transport, and many of the peoples of this region have traditionally pursued nomadic lifestyles based around animal husbandry and the exploitation of wool and hair for making textiles. The nomadic lifestyle necessitates portable looms that can be easily taken down and set up again in new locations. Related terrains and associated nomadic lifestyles extend from the Middle East into Egypt and North Africa. Even though Egypt is part of Africa, some of the earliest evidence for Central and West Asian looms comes from there, so these archaeological looms are reviewed in this section.

There is evidence in Central and West Asia for woven textiles that date back at least 7,000 years, such as from the sites of Jarmo (Iraq), Nahal Hemar (Israel), as well as Cayonu and Çatal Hüyük in modern Turkey. At Çatal Hüyük, for example, archaeologists also found traces of a wooden loom, but it had been burnt and there is no indication of the exact type of loom. Over the following centuries there is more certain evidence for the existence of looms. These can be divided into two basic types, namely horizontal (ground) and vertical (upright) forms. These two forms can then be sub-divided into temporary and fixed versions.

14.1 Horizontal Looms

The earliest form of loom from Central and West Asia appears to be the horizontal loom and in particular the ground loom (cat. 37 and 38). This loom is characterized by a warp beam and a cloth beam that are suspended just above the ground, usually fixed with stakes in the ground. Within this basic type there are many variations, such as temporary and fixed forms, including pit looms and raised types.

14.2 Portable Ground Looms

The oldest known representation of a ground loom dates from the early 4th millennium BCE and originates from Badari, in Egypt (fig. 14.1). The image in question shows a ground loom with two beams that is fixed in place with four wooden pegs. A more detailed depiction of such a ground

Fig. 14.1 The Badari bowl with a painted image of a horizontal ground loom. Petrie Museum at University College London, UC 9547.

Facing page: detail of a silk sash with rows of flowers within ogivals, supplementary-weft patterning. Persia, late 19th century or early 20th century.

loom comes from the Middle Egyptian tomb of Khnumhotep, Beni Hasan, which dates to around 1900 BCE (fig. 14.2). It depicts a weaver and an assistant working at what appears at first sight to be a vertical loom, but is actually a ground loom rendered according to ancient Egyptian artistic conventions. Similar looms can be found in the model from the tomb of Meketre, which dates to the 11th Dynasty (c. 2000 BCE), now in the Egyptian Museum, Cairo (acc. no. JE 46723). In each case the loom is tensioned by four wooden pegs at each corner and has two beams. It also has one or two heddles that were raised by hand.

The weaver would be responsible for passing the weft thread through the shed and for using a heavy wooden beater to make a compact cloth. It was the job of the assistant to open shed and counter-shed, either by raising and lowering the heddle(s), or by moving a shed bar behind the main heddle to the left or right, as contemporary Bedouin weavers do. They would work at the cloth beam end of the loom and to the right of the cloth being produced. The cloths produced were mainly plain weaves, and tended to be warp-faced.

This type of loom continues to be used to the present day and can be found across North Africa, in the Mediterranean region (section 19.2) Jordan, as well as further east, in Iran, Turkmenistan, and Uzbekistan, and as far east as the Qinghai-Tibet Plateau. It is particularly associated with nomadic and semi-nomadic groups, such as the Bedouin (cat. 38) and various Turkmen peoples, as it can easily be rolled up and moved elsewhere when needed. By the 20th century, instead of wooden warp and cloth beams, metal bars (including scaffolding bars) were often used. In most cases it is women weavers who use this type of loom.

14.3 Fixed Ground Looms

By the 1st century CE various settled groups had started to use fixed ground looms that were associated with one place. There is some archaeological evidence for this type of loom in West Asia that dates to the early 1st millennium CE, such as the pit found at the Monastery of Epiphanies at Thebes and that from the site of Dayr Abu Hinnis.

Sometimes the weaver sat on a stool in front of the loom. On other occasions, pits were used, whereby the weaver sat at the same level as the web, with legs and feet resting in a pit underneath the loom (figs. 14.3–14.6). Eventually foot treadles were introduced to the pit looms, which made raising and lowering the heddles easier and quicker. Photographs from early 20th century Iraq, Syria and Turkmenistan show weavers using pit looms. Sometimes these looms were used outdoors; on other occasions they were constructed indoors. Some women weavers in Iran still use a pit loom to produce a type of cotton textile called *kaabafi*.

Another significant development was the introduction of raised looms, which have treadles and a pick-up system for the heddles. This type of loom was used in medieval Egypt and Syria, for example, and used large stones for tensioning the warp threads rather than a fixed warp beam. Versions of this type of loom were also used throughout Central and Western Asia and parts of India and continue to be used to the present day (figs. 14.3–14.6, cat. 34).

Fig. 14.2 Depiction of a ground loom from the Middle Egyptian tomb of Khnumhotep, Beni Hasan, c. 1900 BCE (after a facsimile painting by Norman de Garis Davies, Metropolitan Museum of Art, New York. Public domain image).

14. Central and West Asian Looms

Fig. 14.3 Horizontal loom with the weaver sitting in a pit. The warp threads are tensioned using stones, Tashkent, c. 1871–1872. Library of Congress. LC-DIG-ppmsca-09955-00018.

Fig. 14.5 Man working at a pit loom that is tensioned with a large stone (not visible), Iraq, 1932. Library of Congress, LC-DIG-matpc-16143.

Fig. 14.4 Pit loom weaver at work, Majdal Palestine, around 1935. Library of Congress, LC-M33- 10967.

Fig. 14.6 Syrian silk weaver working with a treadle loom set in a pit. Mount Lebanon, Syria, 1914. Library of Congress, LC-USZ62-69089.

14.4 Vertical Looms

The earliest vertical looms appear to be simple frames that had the warp threads stretched between the top and lower beams. Sometimes these frames were used for making mats, while other loom forms were capable of making much more complicated woven forms.

As with the horizontal looms, there are ancient Egyptian depictions of these vertical looms. They were used by men and appear to have been introduced in about 1500 BCE. In particular they can be seen in various tomb paintings at the Theban tombs near Luxor (fig. 14.7). In some cases, one man is using the loom, but there are representations of two men working on this type of loom, meaning that a much wider cloth could be woven, up to 1 m or more. Traces of this type of loom were found at the Workmen's village, Amarna in Middle Egypt, which dates to the mid-14th century BCE. Actual examples of the wide cloth that was probably woven on a two-person loom occur among the many textiles found in the tomb of the Egyptian pharaoh, Tutankhamun, who died in about 1323 BCE.

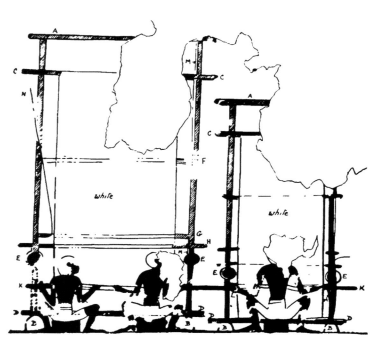

Fig. 14.7 Depiction of a vertical one-man loom and a two-man loom from the Theban tomb of Thutnofer. Tomb 104, c. 13th century BCE.

By the Roman and Coptic Periods in Egypt and the eastern Mediterranean, vertical looms became associated with the weaving of large, draped garments, notably the toga. It has been estimated that looms of up to 5 m in width and 3 m in height were in use to produce the elite forms of togas. Vertical looms on this scale continued to be used in Iran until the beginning of the 21st century, as can be seen with the *zilu* looms in Meybod (section 14.7). They survive in the form of carpet looms that are associated with Turkey and Iran, as well as Afghanistan and Turkmenistan.

14.5 The Egyptian Carpet Loom

Examples of medieval Egyptian carpets are rare, but what has survived indicates that a high level of skill had been achieved by this period. In addition to complete examples in museum and private collections, notably in the Metropolitan Museum of Art, the Victorian and Albert Museum, and the Textile Museum in Washington D.C., fragments of carpets have been identified during a number of excavations, such as those at Fustat (the old capital of Egypt and now part of Cairo), Quseir al-Qadim on the Red Sea coast, as well as far to the south of Egypt, at Qasr Ibrim. These three sites give an indication of the widespread use of piled floor coverings.

By the 20th century a variety of different types of carpets were being made, including those with a wool warp and weft (written wool/wool), cotton/wool, silk/silk, as well as cotton/silk forms. Most carpets now have cotton/wool construction.

The main type of loom associated with Egyptian carpets is the vertical loom, of which cat. 39 is an example.

14.6 Warp-Weighted Looms

There is archaeological evidence for this loom in the regions that are now Jordan, Lebanon and western Syria. Evidence for the use of this type dates back to the Bronze Age, for example the lines of loom weights found at the site of Tell Abu al-Kharaz in Jordan, which date to the 2nd millennium BCE or earlier. This type of loom seems to have disappeared in Southwest Asia by the mid-1st century CE. The warp-weighted loom is discussed in more detail in section 20.4.

14.7 The *Zilu* and the *Zilu* Loom

The modern *zilu* from Iran is a flat woven cloth made from a weft-faced compound plain weave (*taquete*) in two colors, with a design that is double-sided. Both the warp and the weft are made from a thick, cotton yarn. The *zilu* is regarded as a cheapish, but hardwearing floor covering that is used in mosques, hotels, shops, etc. Yet, the history of this type of cloth and the loom it is made on, is much longer than many people realize.

The production of *zilu* style cloth can be traced back to the Sasanian Period (224–651) in Iran. An example of a textile in *taqueté*, in cotton and wool, was found at the Shahr-i Qumis excavations in northern Iran and dates to around the 6th century CE. (Metropolitan Museum of Art acc. no. 69.24.35). Other, more complex examples with unknown provenances are housed in various museum collections and are also made from a combination of cotton warps, with wool wefts and cotton wefts. These can be in either plain weave or sometimes twill weave. They are all characterized by compound weave structure and with a pattern that is double-sided. These Sasanian Period examples are often made out of red dyed wool with designs depicting paired birds flanking trees.

By the Medieval Period in Iran, more and more *zilu*-type textiles were being made with geometric designs. An example of this type dating to the mid-13th century was excavated at the Egyptian site of Quseir al-Qadim. It was made using a cotton warp with cotton and wool wefts. Another example, also in cotton and wool, was identified during the 1980 excavations at the old capital city of Egypt, Fustat. Extant examples of *zilu* textiles from the 19th century are made with a thick cotton thread in both the warp and the weft and being produced primarily for floor coverings. These are essentially identical with modern examples.

Two major centers for the production of *zilu* floor coverings are Meybod and Ardakan, both of which lie near Yazd in central Iran. There are two 16th-century examples of *zilu*- type cloth in cotton that include the word "Meybodi," which suggests that the town of Meybod has been producing these forms of textiles for at least 500 years. In the mid-20th century it is said that over 500 families were weaving *zilu* in Meybod. By 1981 there were 40 weavers and at the beginning of the 21st century only a handful of weavers were still active (fig. 14.8, cat. 40).[1]

Notes

1 For futher reading on the *zilu* and the *zilu* looms, see Beattie 1981, Vogelsang-Eastwood 1988, and Afshar 1992.

Fig. 14.8 A weaver using one of the last *zilu* looms in operation in Meybod (photo: Gillian Vogelsang-Eastwood).

15. Traditional Weaving of Uzbekistan

Binafsha Nodir

To understand loom development in Uzbekistan, it is necessary to consider the traditional ways of life of the peoples of the Central Asian Interfluve, formerly known as Transoxiana.[1] Before the 20th century the two main ways of life revolved around sedentary agricultural production and nomadic herding. The cultural separation between the sedentary and nomadic peoples was most evident in their traditional dwellings. Sedentary peoples built permanent homes of mud bricks and fired clay, while nomads used portable dwellings, particularly the yurt, which is uniquely adapted to a shifting lifestyle centered on pasturing cattle, sheep and camels. There was a corresponding difference their craft productions. Sedentary urban dwellers made textiles from cotton, silk, and wool, such as pile rugs that were woven on large vertical looms and textiles woven on horizontal looms. Nomads, whose main raw materials were sheep and camel wool, made wool products and fabrics on portable looms for decorating yurts, clothing, and horse and camel trappings. Despite the simplicity of their loom, craftswomen managed to produce remarkable textiles, mainly flatwoven, with skillful and original designs.

15.1 The Origins of Weaving in Uzbekistan

The date when looms first appeared in Uzbekistan is unknown. Production of fabrics from cotton, and silk thread, on the territory of Uzbekistan in ancient times is indirectly testified by archaeological finds of an ivory object in the shape of a human hand for unwinding silkworm cocoon threads (fig. 15.1),[2] and wall paintings depicting local people in colorful clothes.[3]

Fig. 15.1 Ivory and bronze object, used to unwind cocoon thread. Khalchayan, the beginning of Christian Era (photo: Aleksandr Shepelin).

Facing page: semi-silk ikat fabric woven by Margilan master weaver, Rasuljon Mirzaakhmedov for the Oscar de la Renta collection, 2004 (photo: Nabi Utarbekov).

A World of Looms: Weaving Technology and Textile Arts

Fig. 15.2 Above: painting from the western wall of the main hall of Afrasiab Palace, 7th century. Below: reconstruction (photos: Museum of the History of Samarkand).

More detailed information about weaving production in Uzbekistan appears from the early Middle Ages onwards (5th–8th centuries). Wall paintings found at fortifications at Balalyktepa, Afrasiab, Varakhsha, and others, as well as at Penjikent fortifications in Tajikistan, convey scenes of palace receptions, feasts, and royal hunting, with people dressed in richly decorated fabrics[4] (fig. 15.2). These evidences show that fabrics of high technical and artistic quality have existed in Central Asia since antiquity and the early Middle Ages. These fabrics were highly valued not only in the West, but also in China, where they were exported.[5] There have been numerous publications on pre-Islamic weaving in the region, but many questions remain, such as the design of Sogdian looms, the composition of dyes, and the patterning techniques.

15.2 The Islamic Period

By the early Islamic Period, weaving was already a rich tradition. The term *zandanechi*, meaning cotton fabric, is frequently mentioned in Islamic written sources in relation to weaving production in Central Asia in the 10th century. These fabrics were of such high quality that they were a sought-after item of export. Narshahi, mentioning Zandana village near Bukhara, wrote:

"So-called *zandanechi* are exported from here. They are so named because they are produced in this village. The fabric is good and at the same time it is produced in large quantities. In many villages of Bukhara, the same fabric is woven and called *zandanechi*, because inhabitants of this village began to manufacture it the first of all. They export the fabric to all areas: to Iraq, Fars, Kirman, Industan and others. All grandees and kings sew clothes from it and buy it at the same price as brocade."[6]

Among the branches of handicraft industry in Central Asia from the late 7th century to the beginning of the 13th century, weaving was one of the most important and widespread. Prior to the Mongol invasion, the main production centers of silk and cotton fabrics, as well as wool fabrics, were Merv, Bukhara and Bukhara oasis, Samarkand, Rabinjan and its district, Kyat in Khorezm, as well as Termez and Chaganian. The major centers for producing fabrics for export, were commercial and industrial villages such as Zandana, Vardan, Iskidzhkat, and Dabusiya, located in the districts of Bukhara and Samarkand.[7]

The high level of weaving art in Bukhara and Merv is evidenced by the descriptions of medieval authors describing such weaving workshops as *beit at-tiraz*.[8] The products, including various fabrics and carpets, entered the Caliph's treasury in the form of tribute due to their high quality. During this period, there was a noticeable increase in Turkic influence in all spheres of life, especially in carpet weaving, which harks back to nomadic traditions. The production of silk and semi-silk fabrics was traditionally the role of urban producers, though little evidence remains of the details of production.

The variety of types of products and ornamentation of weaving products of the 14th and 15th centuries can be seen on the miniatures created in the cultural centers of the Timurid empire—Samarkand, Bukhara, Herat, Tabriz, etc., though neither images of weavers nor looms have been found.

15.3 The Modern Period

In the second half of the 19th century, after the conquest of Central Asia by Tsarist Russia, many specialists visited the region and studied the ethnic cultures. Their studies left many drawings and sketches from the period, and the first photos depicting traditional life. In 1865, by order of the Governor-General of Russian Turkestan, Konstantin Petrovich von Kaufman, a unique photo album was created, called the "Turkestan Album," which is an important visual source for the weaving arts of the region. The "Trades" section in the album contains images of masters preparing yarn, sellers of fabrics and carpets, and traditional looms and weaving processes.[9]

In the middle part of the 20th century there were numerous studies by ethnographers and art historians that provide information about the production of carpets and fabrics, as well as description of looms. In this regard, the works by B. Karmysheva in Tajikistan and Uzbekistan and O. Sukhareva in Uzbekistan are of special interest. Important documentation for carpet weaving and technological processes in Central Asia comes from the research carried out by G. Moshkova,[10] which focused on weaving in Uzbekistan. B. Karmysheva noted:

"… semi-nomadic Uzbek tribes… were characterized by weavings made from wool on hand looms. Even the rather thin fabrics used for sewing of clothes, blankets, tablecloths and various small items, as well as pile carpets and silk fabrics that were found occasionally were all made on hand looms. In cases where the production of cotton fabrics on a treadle loom was encountered among the Uzbek tribes, this indicated a shift

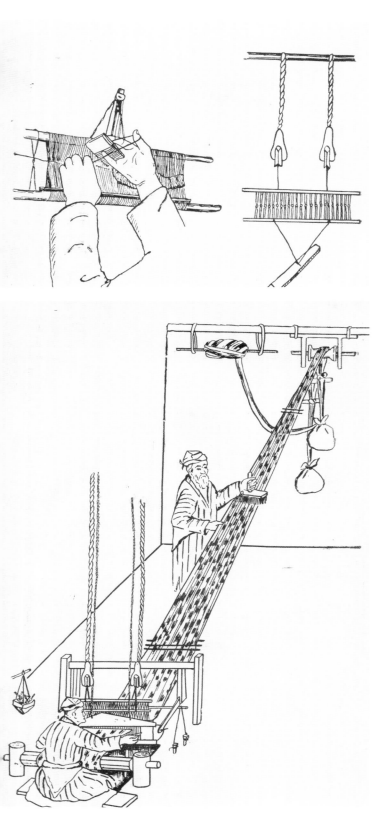

Fig. 15.3 Above left: tying of heddles. Above right: detail of a heddle loom (after Sukhareva 1962). Below: a weaver and a reeder of the warp in a weaving workshop in Bukhara.

towards settled character and mixing with a sedentary population... production of fabrics in the settled population, the Tadzhiks and Uzbeks-Chagatais, is characterized first of all by weaving on a horizontal loom with treadles... As for production of coarse wool fabrics and bedding, they use a narrow handloom of the same type as that widely used among the nomadic and semi-nomadic populations of Central Asia."[11]

The loom for producing silk textiles in Bukhara in the late 19th to early 20th centuries was described in detail in O. Sukhareva's writing:

"The loom for silk was rather complex device, consisting of several separate parts operated together. The opening of a shed by simply pressing treadles and the invention of the flying shuttle were great technical achievements in the mechanization of the weaving process. The loom was set in motion by the hands and feet; the shed was opened by pressing the treadle, the weft was beaten-in with a reed fixed in a frame, the fabric was wound on the cloth-beam by turning a stick, acting as a lever. A well-polished bone shuttle was passed swiftly through the shed. Masters were able to prepare and mount the warp for 100–200 m of fabric. However, even on such machine, it was impossible to weave fabric in a single piece, since the cloth beam, mounted on low supports and almost sitting on the weaver's lap, could not hold sufficient fabric. Therefore, the fabric was cut out from the machine in separate pieces; two such pieces (*dzhura*) constituted one unit for making clothing—*yak libos*"[12] (fig. 15.3).

In the early 20th century in Bukhara, a loom of the typical type for Central Asia was used to produce almost all fabrics. Its consisted of heddles (*gula*) lowered by treadle, the number of which was determined by the type of fabric to be woven, a reed (*tet*) fixed in a movable wooden frame (*dastag*) and a beam (*navard*). The machine was installed on poles or logs dug into the ground, slightly above ground level; treadles (*pomishol*) were located in a pit (*chakhcha*), on the edges of which a board was laid, which served as a seat for the weaver. The Bukhara loom was distinguished by a weighted reed-frame, filled with lead. This was necessary because particularly dense fabrics were produced in Bukhara, requiring firm beating-in of the weft. The nature of the weaving was determined by the number of heddles and the warp setup. Double-sided fabrics in plain weave (*alocha, bekasab, adras*) were woven on two heddles with two treadles; more dense fabric kanaus, although it had the same interlace, was woven on four heddles, connected in pairs with two treadles.

The most complex technique was reserved for the weaving of velvet. It required two kinds of warp, one for the silk pile, composing the pattern *abr*, the other made of single-color untwisted silk (*homak-bershum*) or cotton, securing the weft. There were two warp beams too: a silk warp beam (*pesh-navard*), tensioned with small weights (1–3 kg). The other warp was attached to the second beam (*pas-navard*), tensioned with two heavier weights of 10 kg each. The machine had six heddles and five treadles for controlling the warps. The velvet pile design was produced by wrapping loops of warp around thin brass rods inserted into the warp, after which the weaver cut the loops with a sharp knife or razor (*poku*), The rods had a small groove to guide the cut and ensure its smoothness (fig. 15.4).

Fig. 15.4 Velvet weaving (*a'lo bakhmal*) in Fergana Valley, 2011.

Above: the weaver is inserting metal rods between threads.

Left: the weaver cuts warp loops with a special knife, releasing the pile (photos: Christine Martens).

By the mid-20th century handmade production of various silk and semi-silk fabrics became widespread and economically important in Uzbekistan. The subsequent colonization of Turkestan by the Russian Empire had a negative impact on the weaving industry, however. Local craftspeople were unable to withstand competition from imported factory products from Tsarist Russia. In order to reduce cost and save time, they began to use factory materials and new raw materials, replacing natural dyes with artificial dyes, which adversely affected the quality of textiles. A further, devastating blow was dealt by economic and ideological reforms under Soviet rule, when private property was liquidated. As a result, artisans were deprived of the right to carry out independent and individual economic activities, and those who tried to work as before were prosecuted by the authorities for illegal entrepreneurship.

By the middle of the 20th century, the consequences of this disastrous decision became apparent. Craft traditions that had been passed from generation to generation began to disappear. Work in cooperatives in the 1930s and 1940s turned masters into lackeys; they could not sell their own products and lost the right to free, individual creation. Competition and respect for individual skill also disappeared or became devalued. The traditions of the past began to be forgotten. Artificial aniline dyes came to replace vegetable dyes, while machine production destroyed the very essence of the centuries-old art of weaving.

15.4 Weaving in the Post-Independence Period

With the independence of Uzbekistan, master weavers set themselves the task of reviving the traditions of Uzbek *abra* fabrics. Thanks to efforts in the early 1990s, recipes for vegetable dyes were revived, along with the production of ancient types of fabrics and ornamental silk weaving. Once again in the traditional residential neighborhoods—*makhallas*—workshops for handweaving silk appeared, though this revival has been limited to the Fergana Valley. In Surkhandarya region, in Baysun, the Center for Handicraft has revived the technology of production of the unique cotton fabric *dzhanda*. Today more than 30 workshops are engaged in silk processing in Margilan, and around a hundred craftsmen pattern and dye fabrics, two

Fig. 15.5 Loom, Margilan (photo: State archive of the Fergana region).

hundred are involved in the preparation of yarns, and more than a thousand are directly involved in weaving. The basic technology has not changed much. For weaving silk and semi-silk fabrics, the same *dukon* looms are used, as in the old days, with minor changes (figs. 15.5 and 14.13). Looms are now made of pine, where previously walnut was used. They use the same basic design for all types of fabrics, differing only in the number of heddles and size of the frame, which determines the width of the fabric. Previously, the rules did not allow a deviation from the accepted width of 24–30 cm, but today the fabric width can be up to 50 cm.

15.5 The Looms and Weaving Process of Silk Ikat *Abra* Fabrics

The making traditional silk ikat *abra* fabrics is laborious and involves several stages. Warps of 200–300 m long are prepared on a special machine in small bundles—*libits*. The number of threads in each *libit* (40 or 60) depends on the width of the fabric and its density. *Libits* are connected in pairs and wound on a frame. The distance between the bars on the frame is defined by the length of the design repeat, ranging from 1.4–2.25 m. This

Fig. 15.7 Late master *davrakash* Inom Madorov (d. 2018), distributing silk warps into *libits* using a *davra*, Margilan, 2012 (photo: Elzara Muzaffarova).

process was carried out by the master *davrakash* (figs. 15.6 and 15.7).

On the surface formed by the *libits*, the pattern artist (*chizmakash*) sketches half of the pattern on the vertical axis using a thin stick dipped in ink made from soot, marking its contours with dashes (fig. 15.8). Knowing dozens of traditional patterns by heart, the master does not use either patterns or stencils. Sketches are only used for creating new designs. Areas of color are laid out according to the principle of progressive dyeing of the pattern in different colors, and details of the pattern are alternately reserved by the master *(abrband)* by tying individual parts of the *libits* with cotton threads (fig. 15.9). These days sticky tape is often used instead of thread.

Fig. 15.6 The process of distributing silk warps into small bundles (*libits*) on a special frame called *davra* (photo: Rasul Mirzaahmedov).

Dyeing is carried out by immersion of prepared *libits* in boiling dye baths for several minutes. When creating multicolor patterns, the dyeing of individual sections is carried out in a strict order. For dyeing yarns, both natural and artificial dyes are used, including onion peel, madder (*ruyan*), yellow larkspur (*isparak*), pomegranate peel (*anor pusti*), flowers of Sophora (*tuhumak*), mallow (*gulkhayri*), and imported indigo and cochineal (fig. 15.10).

After all the colors have been applied and fixed, the warp is freed from its ties and dried, and the unwinding and unfolding of paired *libits* is carried out to reveal the full design (fig. 15.11). The warps are mounted on the loom (cat. 41) and the master weaver then takes over. The final stage of the process is glazing the finished fabric using egg white (*kudunglash*) and beating with a wooden hammer (*kudung*) to produce a glossy surface (fig. 15.12). The result is a brilliant and colorful textile with a unique appearance (fig. 15.13).

Fig. 15.9 Late master *abrband* Turgunbay Mirzaakhmedov (d. 2006), tying warp threads, Margilan (photo: Christine Martens).

Fig. 15.10 Master *abrband* Rasuljon Mirzaahmedov and his son dyeing warp threads in indigo. Handicraft Development Center, Margilan, 2017 (photo: Nabi Utarbekov).

Fig. 15.8 Pattern artist (*chizmakash*) applies a pattern on the warp threads for satin, Kokand, 1931 (photo: Photography Foundation of the Samarkand Museum-Reserve).

Fig. 15.11 *Ochdi-kushdi*, the process of unwinding and unfolding of the pattern of the warp threads, Margilan, 2012 (photo: Elzara Muzaffarova).

Fig. 15.12 Master *abrband* Rasuljon Mirzaahmedov glazing the finished fabric using a solution of egg white, Margilan, 2017 (photo: Nabi Utarbekov).

Fig. 15.13 Semi-silk fabric *a'lo bakhmal*, woven in 2010 by Rasuljon Mirzaakhmedov with the planet *sayoralar* pattern (photo: Nabi Utarbekov).

Notes

1 In antiquity the interfluve of Amu Darya and Syr Darya was called Transoxiana, in the Islamic Period it was called Maveraunnahr and now this region constitutes the territory of the Republic of Uzbekistan.

2 Pugachenkova 1991, 301.

3 Pugachenkova 1991, 301.

4 Al'baum 1975, 15–110.

5 Bichurin 1950, 310.

6 Narshahi 1897, 23.

7 Belenitsky, Bentovich and Bolshakov 1973, 270–274.

8 Narshahi 1897, 29.

9 The complete set of all volumes and parts of this album is stored today only in three collections: in the National Library of Uzbekistan named after A. Navoi in Tashkent, in the Russian State Library in Moscow and in the Library of the US Congress in Washington D.C.

10 Moshkova 1970.

11 Karmysheva 1979.

12 Sukhareva 1962, 141.

16. Voice of the Weaver: Margilan and Ardakan

Rasuljon Mirzaahmedov, born in 1973, represents the seventh generation of ikat weavers in Margilan city—famous for silk production—in Central Asia (figs. 16.1 and 16.2). He apprenticed under his father, Turghunbay Mirzaahmedov, a master of ikat weaving in Uzbekistan. Mirzaahmedov's ikat creations, *a'lo bakhmal*, and Bukhara *adras* fabrics, were used in the modern fashion collection of the American designer Oscar de La Renta. Rasuljon's work have revived the local and international interest in these historic textiles. In 2005, he won the 2005 UNESCO award "A Seal of Excellence" and Crafts Prize for the Asia-Pacific region.

I am 45 years old, the ninth of generations of weavers. I started at an early age with my father, a famous master weaver, Turgunbay Mirzaakhmedov. I have three children, two sons and one daughter. They are the 10th generation of weavers. I love to create new patterns; I like this process, painting, drawing, and sketching... I use traditional-style elements, made the same way as [those by] my father and grandfather... 90 percent is traditional style, 10 percent modern. It is a great pleasure to be invited [to the exhibition A World of Looms *] and share my work and see other master weavers from all over the world, and I am glad that my loom will be in the museum [collection].*

Fig. 16.1 Rasuljon Mirzaahmedov demonstrating weaving at NSM (photo: Elena Phipps).

Fig. 16.2 Rasuljon Mirzaahmedov (photo: Tatyana Trudolyubov).

Rastjouy Ardakani Mirzamohammad, age 64 (fig. 16.3), is a master weaver from Ardakan city, Yazd province, Iran. He is the last generation of *zilu* weavers. He has 8 children, all of whom received university education and do not continue the arts of *zilu* weaving.

I began to learn weaving from my father when I was 7 years old, and I was fully trained by 30 years old. I can weave any pattern that you want. I can also build a loom like this one [in the exhibition]. It takes a few days, depending on whether you have someone to help you.

This work has been passed from generation to generation in my family, and I want to keep tradition going. But in my family, nobody wants to learn to weave or to keep weaving, they prefer to study. The government doesn't support me either, and there are problems with the economy. In fact, I love to work, I love weaving, but I can't make much money from it. It is time for me to give up weaving and retire. I really should teach somebody from the younger generation, but I cannot.

Fig. 16.3 Rastjouy Ardakani Mirzamohammad demonstrating weaving at NSM (photo: Nobuko Kajitani).

Africa

17. African Looms
    Malika Kraamer

18. Madagascan Looms
    Andrée Etheve

19. Voice of the Weaver: Analamanga, Ambalavao, and Tema
    Safidy Raharivony, Arline Ravao, Bob Dennis

17. African Looms

Malika Kraamer

An astounding array of fabrics and looms can be found throughout Africa, the second largest continent after Asia and larger in terms of land area than China, India, United States, and Europe combined. Looms can be found all over Africa and many cloths are still handwoven in huge quantities. These fabrics often have high prestige value (fig. 17.1). Extensive trade in these textiles in and beyond this vast continent has taken place over thousands of years.

Cloth in Africa, as elsewhere, has always been more than merely functional: fashion consciousness has a long history, as attested, for instance, by the contents of Egyptian tombs and (much later) in documents from European merchants trading around the West African coast since the end of the 15th century. Their remarks show how they struggled to keep up with changing preferences for particular styles and types of cloths that they took from Europe and Asia, as well as from one West African port to another to barter for cargo, including enslaved people.

Sources for understanding textile and loom histories are uneven for different parts of the continent and our understanding of past loom technologies is correspondingly patchy. Research funding and attention have historically been much more extensive for Egypt than any other part of the continent and correspondingly more is known about this region than others. For some parts of Africa, available pre-17th century sources are limited and archaeological data often scarce on the topic of weaving, due to lack of funding and research as well as basic preservation issues in tropical climates. In West Africa, for instance, much more research has been undertaken in Mali and Nigeria than in other parts for often idiosyncratic reasons, and the illicit trade in African material has destroyed some of our understanding of the past.

Africa is the world's oldest inhabited territory. In Egypt, weaving can be traced back to at least pre-dynastic times (before 4000 BCE) and this might be true for other areas as well. In West Africa, in present-day Nigeria and Mali, the oldest archaeological evidence of cloth production dates to the 800s and 900s, respectively, with indirect evidence for styles of clothing depicted on terracotta sculptures. It is clear that cotton textile production arose in at least two centers before the 15th century, a western center around the upper

Fig. 17.1 Mama Sebeso II (left), Queen Mother, at the 1999 Agbomevorzã, "Handwoven Textile Festival," wearing a rayon *kente* created in the late 1990s with nonfigurative weft-float designs. *Kente* are examples of high status cloths in Ghanaian society (photo: Malika Kraamer).

Facing page: fabric made of multiple narrow-woven strips (the warp direction is shown horizontal), striped patterns and geometric designs in supplementary-weft patterning. Ghana, early 20th century.

Niger, Gambia, and Senegal watershed, to the edge of the desert, and an eastern center around Lake Chad and the area of the early Hausa Kingdoms.

The African region is often divided in North, West, East, Central and South Africa and consists currently of 54 countries. Most countries host a large diversity of cultures and languages. Weaving traditions constantly change over time; many are dynamic and alive, and of ancient origin. Some technologies have disappeared; many others have not been fully documented. Consequently, this review only considers the better-known types.

Looms facilitate the weaving process in two ways: they keep the warp in tension and, in most cases, provide a shedding device. A few peoples, such as the Tuareg of the Sahara, make cloth by manipulating weft and warp with their fingers alone. In most cases the loom does not determine the design of the finished garment (in contrast to some South American weaving, for example). Textile design depends rather on the nature and color of the fibers, the relationship between warp and weft, and on post-weaving decoration such as embroidery, as well as the design of finished garments.

Many classificatory accounts of looms in Africa over the last thirty years follow the overview given by John Picton and John Mack in *African Textiles*,[1] in which a basic distinction is made according to shedding devices, dividing looms into the two broad categories of single-heddle and double-heddle looms. This distinction will be followed in this review. Another useful study, upon which this overview draws, is the work of Colleen E. Kriger and her historical research on textile production in Africa, in particular the lower Niger Delta in present-day Nigeria.[2]

Looms can also be classified by, for instance, the warping method and how the warp is fixed, such as the continuous-warp loom, the ground loom, and the vertical upright loom, as well as the fibers used or attributes of the cloth woven, for instance the raphia loom, the narrow-strip loom, or other physical attributes or the loom's geographic location, such as the Ethiopian pit loom, Sierra Leonean tripod loom and the Egyptian vertical and horizontal two-beam looms. Local terminology for looms and weaving technology, which has barely been examined and remains a topic for future research, may provide other ways of classifying looms, as well as shedding more light on connections between different textile traditions.

17.1 Raw Materials, and Woven Versus Non-Woven Cloth

Many non-woven materials such as skins and bark cloth were used in Africa, and these may in fact represent the earliest types of cloth. For making woven cloth, a wide variety of fibers are employed, including bast, leaf fibers (such as raphia), cotton, silk and wool. Man-made fibers have gradually gained importance after the late 19th century.

Wool has frequently been used in North Africa, in the Sudan, and the Sahel zones. Berber women in North Africa, for example, use wool for their weft-faced textiles. Silk also has a long history in Africa, but has more restricted uses than wool. Nigerian wild silk and (waste) silk from Europe arriving in northern Nigeria through the trans-Saharan trade were used for luxurious textiles. The domesticated silk moth is only found in Madagascar, though imported silks have been known across the continent since the 17th century, if not earlier.

Little is known about the production of cotton before the 11th century CE, but by that time West Africa had become one of the major world centers of cotton cultivation. Earlier cotton textiles have been found in the Sudan, especially from the late Meriotic Period (200–500). Cotton has also been extensively used in Ethiopia, Somalia, and all the way down to the south of Zimbabwe.

The use of bast fibers was important in the past, but has become uncommon today. The earliest evidence of bast fibers in West Africa comes from excavations in Igbo-Ukwe in Nigeria from the 9th century. Raphia fiber, a type of leaf fiber, also has a long history in Africa, particularly in West and Central Africa and in Madagascar.

17.2 Single-Heddle Looms

A widespread method in Africa for keeping the warp in tension is to stretch it between two parallel beams, on which the warp is either wound continuously or attached in single lengths. All of these looms use a shed stick to retain the natural shed, and usually a single heddle to which half of the warp elements are attached that can be pulled to make the counter-shed.

The beams can be pegged to the ground, in which case the loom is called a ground loom, or else mounted in a frame. Warps mounted in a horizontal plane, pegged to the ground or on a frame, can be found scattered throughout the continent, including Madagascar (section 18), North and West Africa, and were probably once even more widespread than they are now. It was the most common type of loom in East Africa, where it nevertheless disappeared in the early 1900s following the importation of trade cloths. The horizontal Berber loom used to weave tent-cloths was probably introduced as a result of contact with nomadic peoples from further east (section 14.3), while the vertically-mounted Berber loom has different origins and is used for the production of blankets and rugs, as well as male cloaks and female clothing.

Aside from Berber North Africa, looms with frames oriented vertically are mainly found in Sierra Leone, Nigeria, Cameroun, and the Democratic Republic of the Congo. In Nigeria, the vertically mounted loom with one heddle is often placed against a wall and has been used by many different groups. This loom is described in more detail below. The broad width of the cloths woven on these looms meant that one, two, or three pieces sewn together were enough to form a cloth wrapper. The use of this loom diminished after the mid-20th century, as the hand weaving of textiles for everyday domestic use declined in the face of imports. Only those traditions that make cloth for prestige or ceremonial contexts are still flourishing today.

In the forested areas of West and Central Africa, which have been populated for several thousands of years at least, weaving on looms was probably preceded by the use of non-woven fabrics. There is no concrete historical evidence as to when weaving on looms started, but the relatively simple single-heddle loom is probably the oldest type. These looms were used to weave raphia and bast fiber in first instance, such as the raphia loom from the Congo described below. Raphia weaving on both single-heddle and double-heddle looms continues in Central Africa and some parts of East Africa, but the use of cotton and man-made fibers has more or less replaced the use of bast and leaf fibers in West Africa.

Single-heddle looms can also be found in the Middle East, but connections between these types and African looms are speculative, except in the case of Egypt, where Middle Eastern looms are known to have been introduced during the period of the Mamluk Empire, no later than the early 1500s.

17.3 Double-Heddle Looms

Another way to apply tension to a warp is to fix one end of the warp to a framework and the other end either to a weight, as is done in many parts of West Africa, or to a peg in the ground, as in Northeast Africa, including Somalia, southern Egypt, the Sudan, and Ethiopia. On these looms the warp is always mounted more or less horizontally. The shedding device on all of these looms is formed by at least one pair of heddles, hence their classification as double-heddle looms. The heddles are connected by a pulley suspended from above and are operated by foot treadles (fig. 17.2). As with Middle Eastern double-heddle looms and many Asian looms the heddles are of the bidirectional (clasped) type, constructed between two parallel sticks with a set of interlinked loops of string (or other flexible materials) in between, with the warp yarns passing through an "eye" in the center. As these looms are invariably operated by treadles, they are also referred to as treadle looms. The weaver sits behind the cloth beam on which the woven cloth is rolled, using a shuttle to insert the weft and a reed or metal beater to beat in the weft. These looms leave the weaver's hands free to throw the shuttle back and forth, allowing for faster production of cloths, at least in the case of plain weave. As on single-heddle looms, a wide variety of weave structures and patterning are produced.

In West Africa, the warp is usually attached to a drag-stone, a wooden sledge, or a piece of leather or metal upon which a heavy stone is placed. The drag-stone is heavy enough to provide tension for the warp, but at the same time can be pulled towards the loom when the weaver needs to roll the woven cloth around the cloth beam. West African weavers produce long cloths in narrow widths of 15 cm width or less and many meters long, a feature that is unique to this region. Thus, West African double-heddle looms are sometimes called narrow-strip looms. After cutting up the strips at designated places, they are sewn together edge to edge to form a cloth. Paying close attention to putting the correct side up, a distinction that may only be apparent to a trained eye, and lining the strips in such a way that designs match across the cloth is crucial for the aesthetic of the finished cloth, hence this aspect is often the preserve of senior weavers or those having retired from weaving.

Many hypotheses have been put forward on the spread of the double-heddle loom in West Africa and the narrow-strip format, but there is little concrete evidence. The weaving on this loom in the northern savannah in Nigeria predates the spread to the southwest where the production of *aso oke* continues to be a fashionable industry.

In Northeast Africa, the "warp beam" is actually a peg in the ground. The warp is stretched around

Fig. 17.2 Kofi Agbemehia using a double-heddle loom of the coastal six-pole type, with a pair of treadles connected to heddles that open shed and counter-shed for the ground weave. Kpedzakofe, Ghana, 2000 (photo: Malika Kraamer).

a series of pegs and is tied to the last peg. A pit is dug beneath the loom and the weaver sits at the edge with his legs inside so that his feet can operate the treadles. This loom, called a pit loom or pit-treadle loom, is still used in Ethiopia and Somalia, and has an uncertain origin. Some writers suggest that the technology was imported from the Indian subcontinent, as the Ethiopian loom has similarities with pit looms used in North India and the surrounding regions (section 12.2.5).

17.4 Weaving Techniques and Decorative Techniques

Most weaving traditions are either predominantly warp-faced or weft-faced; in only a few areas, such as in the main weaving centers in southern Ghana and Togo, do weavers produce both weft- and warp-faced textiles. In Ghana, weavers make cloths with areas of alternating weft-faced and warp-faced plain weave, often with supplementary designs in warp or weft. This technique has been used at least since the 18th century and employs two pairs of heddles (fig. 17.3). In many warp-faced traditions, the key to creating patterns lies in the way the warp is laid, including the use of a supplementary warp in a few places. In both weft- and warp-faced textile traditions, designs are also created with the ground weft and through the use of supplementary wefts. Openwork is less common, but can be found in Nigeria in textiles produced on both the single- and double-heddle looms. Ikat as a technique for decorating fabrics can be found throughout Africa, especially West Africa and Madagascar.

17.5 Male and Female Weavers

In some regions of Africa, such as Madagascar, Arab North Africa, the Sudan, and Nigeria, weaving has historically been undertaken by both men and women. In others, such as most of West Africa, Ethiopia, East Africa, and Democratic Republic of the Congo, weaving is predominantly done by male weavers. In the case of Berbers in North Africa it is only the women who weave. Throughout the continent, men may or may not be full-time specialists, while most women weave primarily for domestic use, weaving being one of several skills they are expected to have.

In areas where both men and women weave, they mainly do so on different types of looms. In areas with both double- and single-heddle looms, such as parts of Nigeria, Benin, and Togo, men weave on double-heddle looms and women on single-heddle looms. Double-heddle looms seem to have mainly been the preserve of men in most areas, which in the past was often explained in terms of specific taboos against women weaving. Since the mid-20th century at least, some women have taken up weaving on these double-heddle looms, but they

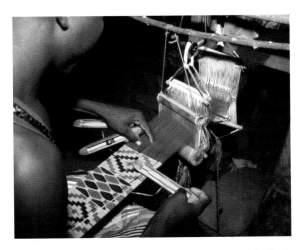

Fig. 17.3 Wisdom Gidiga weaving patterned cloths with two pairs of heddles. Agotime-Afegame, Volta region of Ghana, 2000 (photo: Malika Kraamer).

Fig. 17.4 A. Mortoo weaving using a coastal four-pole portable loom. Anlo-Afiadenyigba, August 2000 (photo: Malika Kraamer).

remain a minority (fig. 17.4). Resistance against women weaving on those looms continues in West Africa. Today both men and women often explain this in terms of expected gender roles; it is just not a female job. Single-heddle looms are in some areas operated by women, like Southwest Nigeria, but in others by men, such as Southeast Nigeria, northern Cameroon and the Democratic Republic of the Congo. There is no single explanation for these gender distinctions.

17.6 Power Looms

Since World War Two, power looms have become widespread in Africa, though the first power-operated looms were already present at the end of the 19th century. The Mungo Mill in South Africa, for instance, has used such looms since 1890.[3]

Fig. 17.5 One of many single-heddle loom variants formerly used in Central African raphia weaving, Congo-Brazzaville, early 20th century (postcard from the collection of Duncan Clarke).

17.7 African Loom Types

The diversity of looms, weaving technologies, and textile traditions in Africa are overwhelming. It is impossible to cover all the loom types in this introductory essay; some of the best-known are described below. Two looms from Ghana are described in the catalog (cat. 42 and 43).

17.7.1 Raphia Looms in the Congo Basin

The raphia looms found in the Congo Basin, which are operated by male weavers, are single-heddle looms employing discrete warp lengths (fig. 17.5). Upright single-heddle looms in Central Africa have numerous variations in the angle of mounting, the absence or presence of a frame, but they all share the use of a discontinuous warp. Single-heddle looms for weaving raphia have also been documented for isolated pockets in West Africa, such as Ghana, Sierra Leone, Southeast Nigeria and Cameroon, and today can still be found in Gabon and Angola. They probably once had a much wider distribution.

On these looms in the Congo Basin, the length of the warp elements, corresponding to single strips of raphia as torn directly from the plant, determines the length of the cloth, which is usually a little over 1 m. Each pick of the weft is also a single length of raphia. Looms using raphia in other parts of Africa, also in areas where these looms have disappeared, did get around the limitation of the single length of a fiber by knotting the fibers together or, in the case of Madagascar, by twisting fibers to create a continuous thread. In 1778, a French trader described raphia wrappers from Benin as being eight or nine feet in length (roughly 2.4–2.7 m).

These looms are usually used to weave plain cloths which can be subsequently embroidered or tie-dyed, but in some areas they are also used to make complex weft patterned cloths created through the use of multiple heddle sticks. Many of the diverse raphia traditions have died out in the middle of the 20th century, but the Kuba people of the Kasai region are still weaving raphia in Central Africa. Raphia weaving was once a central part of both economic and ritual life, but has faded in importance today.

17.7.2 Vertical Single-Heddle Looms in Nigeria

Across much of Nigeria, women use vertical single-heddle looms for domestic weaving. These looms have a long history along the Bight of Benin and Biafra, and its hinterlands. The best-known weaving traditions on this loom are the cloths from the Akwete region in eastern Nigeria and Yoruba textiles in western Nigeria. The many structural variations of this type of loom suggests that different groups made adaptations to these looms at different times and locales, which indicates its antiquity. This loom, as suggested by Colleen Kriger, may be related to the raphia loom of the equatorial rainforest that stretches southeast in the Congo Basin.

The loom is a simple framework that is either freestanding or leaning against a wall. The warp threads are wrapped continuously around the top and bottom beam creating a circular warp. It is much smaller in size than the Berber and Bedouin upright looms. Lashed to a heddle stick, a single heddle of string loops around alternate warps allows the weaver to create a shed by pulling the stick forwards. The counter-shed is created with one or more shed stick(s) that sits between the two groups of warps and is used to pull back the second set of warps. A wooden "sword" is used to beat the weft in tightly, beating downwards. Once as much cloth is woven as conveniently possible, the loop of warps is shifted around the two beams at the front, making another portion available to the weaver. When the entire warp has been woven, except for a few inches, the weaver cuts across the remaining warp threads to remove the cloth from the loom.

In the past, and still in many areas today, married women were expected to spin cotton into yarn and weave it using the vertical loom. Machine woven yarns are also used for this type of weaving, and have been used since the 19th century. The earliest weaving in this vast area was done with indigenous materials, such as raphia and bark, probably on a variant of this vertical loom.

17.7.3 Double-Heddle Tripod Looms in Sierra Leone

Mende weavers in parts of Sierra Leone and Liberia use various tripod and tetrapod looms (fig. 17.6). This is one of the simplest forms of treadle loom.

Fig. 17.6 Tripod loom, Sierra Leone, c. 1900–1910. Vintage postcard (photo: W.S. Johnston, postcard from the collection of Duncan Clark).

The loom derives its name from the stand made from three or four sticks placed over the warp, from which the pair of heddles is suspended. The warp is at one end fixed to a cloth beam, which is supported between two posts and at the other end to a peg or another beam supported by two posts. When the warp is fixed to a peg, a basket sometimes contains the rest of the warp. The beater is not fixed to the tripod, but is supported by the warp elements passing through it. The weaver is able to manipulate the reed beater by a handle, and sits beside the loom to operate the pedals. As weaving progresses, the tripod, heddles and beater are moved along from cloth beam to the warp post. When the exposed part of the warp is fully woven, it is rolled on the cloth beam, a new stretch of warp is exposed (taken from the warp in the basket) and the tripod with heddles and beater is moved back to the cloth beam. The weaving then continues. The biggest difference compared with other West-African double-heddle looms, where the warp is continuously pulled through the heddles, is the fact that the weaver moves the shedding device along the warp. This practice is similar to the fixed single-heddle loom, sometimes also called "raised ground loom," as used in Northeast Africa and Cameroon.

17.7.4 Pit Looms in Ethiopia and Somalia

The key features of the pit loom from Ethiopia and Somalia, like double-heddle looms found in many parts of West Africa (figs. 17.7 and 17.8), are the use of suspended heddles operated by foot pedals, a weighted drag-sled to bring tension in the warp and the weaving of a long strip, usually rather narrow, which is cut up and sewn together edge to edge to form the finished cloth. The treadles are suspended in a hole in the ground and the weaver sits at the edge of the pit with his legs inside so that his feet can operate the treadles. Cotton is the only material used on these looms in Ethiopia and Somalia.

The pit loom in Ethiopia, with a sturdy frame, and its particular depth of the pit and width of the heddles, and therefore consequently of the web of the strips, did not spread to other parts of Africa, nor even to other parts of Ethiopia, at least not until recently.

Notes

1 Picton and Mack 1989.

2 Kriger 2006.

3 See https://www.mungo.co.za/about-us/.

17. African Looms

Fig. 17.7 Man's robe, Hausa people, Nigeria. This robe is composed of narrow strips of indigo-dyed cotton textiles with a check pattern, joined together to make a robe. After tailoring the robe, designs were added by a specialist embroiderer. The robe has been repaired in places with patches of striped indigo textiles. About 2 m x 1.3 m. Private collection.

Fig. 17.8 Facing page: detail of the embroidered designs.

143

18. Madagascan Looms

Andrée Etheve

Madagascar lies off the coast of East Africa and is the fourth largest island in the world. It has a unique culture and history that sets it apart from other regions of Africa. Madagascans speak an Austronesian language, brought to the island by migrants from Borneo around a thousand years ago, and Madagascan weaving today is a blend of East African and Austronesian elements. There are many regional variations of Madagascan textiles and several loom types. The most common type is a local variation of the two-beam ground loom that is found throughout Central Asia, North Africa, and parts of East Africa (Mozambique and Madagascar). It consists of two beams in a frame or secured behind posts, with a single heddle bar that is fixed and held above the warp on pillars. The warp is circular. A rare type of body-tensioned loom that is only found in the far south of the island resembles looms that are used in the eastern archipelago of Indonesia (section 10.1), and was almost certainly brought by migrants. In the past, warp ikat was an important decorative technique in Madagascar, and a few older warp ikat textiles bear a remarkable resemblance to Indonesian ikats. Both the loom and the warp ikat technique are likely to have been brought to the island by Austronesian settlers.

There are many early accounts of textile production in Madagascar. An early 17th century sea captain, Paulo Rodriguez Da Costa, observed that people visited the coast of Madagascar to purchase *rabane* or "fabrics woven with plant fibers that are made with skill and in great quantities." Da Costa further noted that "Few households possess fewer than three of four looms."[1] Robert Everard, writing at the end of the 17th century, gave some clues about the looms used on Madagascar's west coast. He wrote, "In order to produce their fabrics, Madagascan people hammer in four wooden posts, two by two, separated by a distance equal to the half of a length of the cloth that they wish to weave; between two posts. They then place horizontal bars on which the warps are attached."[2] In the 1700s, Robert Ducry, another English explorer, mentioned the abundance of plant fibers, such as raphia, tree bark, and cotton. He observed that silk was also plentiful, "It is harvested on various trees and the cocoons are unreeled by hand."

The basic construction of traditional Madagascan looms today is still the same as that described by P. Heimann in 1930, "The simplicity of Madagascar looms, the 'rudimentary' endless warp, the single heddle shaft that has to be made anew for each weaving, the beater that is taken out of the shed after each weft insertion, the patterns woven with supplementary wefts using a heddle lifted up by hand, mobile loom parts that are rapidly assembled without requiring permanent installation, enable the loom to be used in the smallest houses and the warp system to be rolled up at the end of the day."[3] These conditions remain in place for contemporary weavers, most of whom live in small houses and have limited space to weave. In the coastal areas where people still observe a nomadic way of life, portability is a necessary feature of domestic looms. One needs to be able to easily roll up the warp system and carry it while migrating to a new

Facing page: silk with floral motifs, supplementary-weft patterning, Madagascar, 2018.

area, as well as to keep it safe at night and during periods of bad weather. In the coastal areas in the south of the island people use body-tension looms. On these looms, they weave long colorful patterned bands, *salaka*, which men wear around their waists.

In workshop settings, treadle looms are used. For example, a few small studios—located in the vicinities of silk or raphia producing cities on the road to the east coast—employ treadle looms with four or six treadles to weave fabrics in plain or compound weave in raphia, sisal, cotton, domestic and raw silk. Another type of treadle loom is used in the forest areas in the eastern part of the island by an ethnic group called the *Betsimisaraka* (fig. 18.1). These looms are also equipped with clasped/bidirectional heddles. The heddle shafts are linked to cords passing over a bamboo or wooden beam, pulling one heddle up while lowering the other, alternately. The warp is either attached at the back of the loom or rolled around a warp beam. This type of loom is ideal for weaving fine yarns, but

Fig. 18.1 *Betsimisaraka* weaver making *rabane* (photo: after an old postcard, early 20th century).

it can only produce plain weave. Looms based on the ground loom (fig. 18.2) are more common in domestic settings in Madagascar: three types of these looms are illustrated in the catalog (cat. 43–45).

Some international aid programs have tried to introduce other types of looms such frame looms, treadle looms, and even Jacquard looms. These efforts have met with little success; the new looms are impractical for most weavers because they require more space and have longer warps, which translates to a higher financial commitment each time the loom is set up. Today, the diversity of weaving expression in Madagascar also faces major threats due to the shortage of many raw materials and the requirement for standardized production. The latter is a consequence of emerging markets linked to tourism and export. These factors have encouraged the production of textiles that are quickly woven, have a low warp and weft thread counts, and are almost or completely plain and without any motif.

Fig. 18.2 Weaving on the *akotifahana* loom, Madagascar (photo: Andrée Etheve).

Notes

1 Heidmann 1937.

2 Mollet 1951–1952.

3 Chapus and Dandouao 1951–1952.

19. Voice of the Weaver: Analamanga, Ambalavao, and Tema

Fig. 19.1 Safidy Raharivony demonstrating weaving at NSM (photo: Elena Phipps).

Safidy Raharivony, age 22, was born and raised in Ambodidrabiby Kelifatra, a weaving village cooperative in the Analamanga province, Madagascar. Weaving is part of his family heritage, and he learned to weave from his father. His specialty is weaving the *akotifahana* brocade (fig. 19.1).

The village of Kelifatra is an old cooperative silk village. I was born there. I grew up under my parents' looms: weaving is in my blood and in my tradition. I am happy, calm and completely absorbed in my work when I am weaving. Our life is all around silk: mulberry trees and silk worms, and we prepare the threads on traditional tools like our ancestors, and the result is special.

In the Analamanga province, the deceased are wrapped in cloth called "akotifahana" before entering the family grave. Every Malagasy woman has one or more large silk scarfs with "akotofahana" designs for various family occasions. Silk and akotofahana are very important here in Madagascar: it is a sign of identity. I come from an old family of weavers and have to perpetuate my ancestors' knowledge in silk brocades.

My loom is a traditional loom with a simple frame on legs equipped with extra heddles for the supplementary weft to weave "akotifahana" brocade from Madagascar. As we live away from the towns, we are rather isolated from the direct buyers and we are obliged to sell our work to "revendeurs" in the towns for the most parts of our products.

My dream is to keep alive our work and for that developing our family's silk work, to show it in national and international exhibitions and to continue on this beautiful silk work. I am proud of our traditional work. Participating in the [A World of Looms] exhibition with one of our "akotofahana," weaves to show the public, meeting a lot of weavers from all parts of the world, seeing their looms and their works, and discovering the fabulous textiles of China, all was so amazing! I returned [to Madagascar feeling] proud of all the weavers in the world.

Arline Ravao lives in South Madagascar in a little town called Ambalavao, about 100 km from Fianarantsoa. She is from a well-known family of weavers, and she is an expert on weaving traditional funerary clothes with natural colors (fig. 19.2).

I am from Madagascar. I am 40 years old. I have woven for 20 years. I learned to weave from my family, from mother and grandmother. I also went to school to learn to weave. I like to weave.

Fig. 19.2 Arline Ravao demonstrating weaving at NSM.

Bob Dennis is a weaver from Shrohume, close to Agbozume in the Volta region of Ghana. He currently lives and works in Tema, close to the capital Accra (fig. 19.3).

The most important part to me about the weaving we do is how adaptable this weaving is, and how we are able to play around with motifs on our narrow strip looms, as compared to other looms… Most of the time people think [that] these symbols are embroidered on after the weaving. And we told them, 'No.' We play around [with] all these things with our fingers picking the weft threads, and we come up with all manner of designs, and symbols, and logos that you can imagine. That is one amazing thing about the weaving that we do here in Ghana.

We have a very diversified customer market, in the sense that [some] people are very choosy when it comes to the type of things they want, and some just flow with whatever you have on display or you have in stock for them… All manner of customers, you meet. Some are also very interesting, very innovative. They contribute a lot to the design that we do, both here and abroad. Especially when you meet people who have textile background, you share ideas with them; and that makes your job very interesting because you always want to share expertise and, at least, talk to people about the view of what you do. It encourages you.

This [kente] cloth in Ghana is our traditional cloth… It is something that has been passed on from generation to generation; so, it is more or less like a generational tradition that our fore-fathers bequeathed our grandparents with. Most of these cloths that you see today are traditional; [it is] just that we try to adapt to the new colors that we have now, by changing the traditional colors and the traditional patterns to fit what we do today…

Fig.19.3 Bob Dennis sampling colors for a new design in his shared own workshop in Kpone, Ghana, 28 July 2018 (photo: Malika Kraamer).

Europe

20. Ancient European Looms
    Eva Andersson Strand

21. Voice of the Weaver: Two Ancient Tales
    Magdalena Öhrman

20. Ancient European Looms

Eva Andersson Strand

In studying ancient European looms, textile scholars are fortunate that at least one type—the warp-weighted loom—leaves evidence of its use in the form of loom weights, some of which are even found in rows in their original positions. In addition, this type of loom remarkably survived until recent times in a few rural communities in Scandinavia, as documented by Marta Hoffmann.[1] In contrast, little is known about other types of looms that were used in ancient European societies. Evidence for those looms is scarce since wood, the main construction material for most looms, rarely survives for a long period of time.

When considering loom evidence—or the lack thereof—it is important to examine all types of material remains as well as potential inherent biases of archaeological records. To gain new insights and perspectives, scholars need to combine different approaches and compare the evidence from within and outside Europe. Textual and visual records have revealed important information on looms; various early Egyptian looms, for example, are known only from ancient iconography. Marta Hoffman's ethnographic study of the Scandinavian warp-weighted loom has also proved key to the reconstruction of loom parts from archaeological sites. In other words, in order to achieve a fuller understanding of loom varieties in ancient European societies, one must consider the archaeological, textual, and iconographical evidence, as well as ethnographic data and loom reconstructions. Looms that are usually discussed in the European context are the vertical two-bar loom and the warp-weighted loom, as well as frame looms with treadles. This essay also includes the horizontal two-bar loom because it—being one of the earliest looms in the Middle East—was likely to have been used in southeastern neighboring parts of Europe as well.

20.1 Two-Bar Looms: Horizontal and Vertical

A simple method to create tension on the warp is by attaching the ends to two parallel bars, which can be horizontal or fixed into vertical uprights.[2] The vertical type is relatively rare in Europe, and is known mainly from old illustrations. The history of vertical two-bar looms in Europe and their distributions are poorly understood, though this format has survived until the present day in the form of upright looms that are used for tapestry and carpet weaving.

20.2 Horizontal Ground Looms

The horizontal ground loom is considered to be one of the world's oldest loom types. The earliest known depiction is painted inside a pottery bowl from Badari, Egypt, dated to the early 4th millennium BCE (section 14.2).[3] Early pictorial evidence of this type of loom also comes from the eastern Mediterranean, for example, on a cylinder seal from Susa (in present-day Iran) that dates from the 4th millennium BCE. In addition, a later Egyptian dynastic painting (fig. 14.2) depicts two weavers sitting on either side of the loom: one is changing the shed, the other entering and beating

Facing page: pyramidal loom weights modelled after the archaeological finds from Insula VI.I in Pompeii (photo: Magdalena Öhrman).

the weft. The heddle rod is supported with heddle jacks, a feature that is still seen in present-day looms of this type.[4]

Archaeological textile findings suggest that the ground loom was primarily used for weaving plain weave and basket weaves,[5] mostly in linen and in various degrees of fineness.[6] These findings show that, despite the apparent simplicity of this loom, it can yield textiles with high thread counts in both warp and weft. The average width of the textiles is about 1.2–1.5 m, but there is mention of widths of up to 2.7 m. Regarding textile length it has been suggested that "[t]he length of the cloth woven on a ground loom is limited only to the amount of thread spun."[7] The intended use of the final textile products, however, would also influence the woven dimension.

The horizontal ground loom is still in use today across much of North Africa, the Middle East, and Central Asia, extending east to the Qinghai-Tibet Plateau, China. In Jordan, for example, it is used to produce rugs and other relatively coarse wool textiles. In the Bani Hamida Bedouin weaving project, the width of the loom varies up to about 1 m, while the warp is around 10 m long. When cut this length can be used for four rugs.[8] The technique used is mainly plain weave, but weft twining is also used to make bands as well as rugs. Warping is done directly on the loom, and the heddle, which is of the simplest type that pulls in one direction only, is made at the same time: a convenient and fast process.

20.3 Vertical Two-Bar Looms

There are several types of vertical two-bar looms. All tension the warp between two horizontal beams that are fixed to uprights, forming a rigid frame. The weaving is begun at the base, and the weft is beaten-in downwards. The simplest type consists of a wooden frame without any shed-bars, while more elaborate looms, such as the Egyptian and Roman vertical looms, may have had the means to open several different sheds.[9] The warp can be circular (forming a tube), and in this case its length is dictated by the distance between the two horizontal bars (fig. 20.1). Another way to fasten the warp is by using a twined starting cord attached to a beam, on which the warp is rolled. In this case, the length of the warp may be longer. During weaving, the weavers usually sit in front of the loom, and the number of weavers may vary depending on the width of the cloth to be woven.

It has been suggested that the two-bar loom originated in Syria or Mesopotamia, but its earliest visual representation occurred in Egypt during the last part of the 2nd millennium BCE (section 14.4). It has also been suggested that this loom was developed with the advent of wool spinning technology. Since wool fibers hold dyes easily, these colorful yarns probably inspired weavers to experiments with patterned weaves, such as tapestry and supplementary weft. The two-bar loom is particularly suited to these woven techniques.[10]

Fig. 20.1 Two-bar loom with a tubular warp. Land of Legends open-air museum, Lejre, Denmark (photo: Ulla Mannering).

Textiles woven on a two-bar loom can be very wide and long: a great advantage. However, the warp on this loom is less flexible, allowing only narrow shed to be formed, preventing a long, continuous shot of weft.[11] To produce a wide cloth, therefore, it is likely that several weavers would work simultaneously to complete each weft shot, passing a shuttle from hand to hand. Analyses of early Iron Age bog textiles in Denmark supports the conclusion that two to four weavers were working on the same loom.[12] In the case of the Danish finds the weave is a coarse twill with 10–12 threads per centimeter.

Today, the vertical two-bar loom is used all over the world, generally for carpets and non-continuous patterns/tapestry weaving in plain weave. The looms can vary enormously in size from a table tapestry loom that is 64 cm wide[13] to a tapestry loom with a width of 4 m, such as those used by artists in Atelje 61 in Petrovaradin, Serbia.[14] This loom is also used to weave kelim carpets in Turkey. Here, the loom height is 1.8 m and the width 1.2 m. The length of the warp (fastened to a metal rod attached to the main beam) can reach up to 10 m long, which is enough for three to four carpets.[15]

20.4 Warp-Weighted Looms

A warp-weighted loom is generally upright or positioned at a slight angle to the vertical, leaning against a wall or a roof beam. In contrast to the rigid frame of the vertical two-bar loom, the top beam of the warp-weighted loom often rests loosely in a cleft. The weaving is begun at the top, and the weft is beaten-in upwards. The hanging warp threads are kept taut by the attached loom weights (fig. 20.2), which can be made of either clay or stone and may vary in size and shape in different regions and different time periods. In some areas such as Scandinavia, the basic type of loom weights have remained the same for more than a thousand years; in other areas, such as the Aegean of the Middle and Late Bronze Age, different types of loom weights have been used in different time periods.

Fig. 20.2 The warp-weighted loom (photo: Linda Olofsson © The Vorbasse project and Centre for Textile Research, University of Copenhagen, Denmark).

The functional parameters of the loom weights are their heaviness and thickness. Other factors such as the shape of the weights (round, pyramidal, conical, etc.) may simply be a matter of tradition rather than function.[16] Loom weights in England during the Roman Period, for example, were mainly pyramidal or conical just as they were in other parts of the Roman Empire. During the following Anglo Saxon Period, however, the loom weights in the same region became donut-shaped. The latter is the same type used in Scandinavia, where many migrants to the British Isles originated.[17] As Marta Hoffman has shown in her research, the number of weavers working on a loom varies: one weaver is enough for a narrow loom, but multiple weavers may work together on a larger loom.

Elizabeth Barber has suggested that the warp-weighted loom was already present in Central Europe, for example in Hungary and perhaps Anatolia by the 6th–7th millennium BCE, in the early Neolithic Period. The use of the loom seems to spread westwards and northwards into Greece, northern Italy and Switzerland, and then to Scandinavia and the British Isles during the Bronze Age.[18]

Similar types of loom weights do not guarantee that the looms themselves are the same. Regional variations may exist, for example between the Scandinavian Viking Age and the Aegean Bronze Age—though we have little direct evidence of this. The warp-weighted loom can be used to produce many different types of textiles and is especially suited to producing twills. It is clear that the fabrics produced would vary based on the different societies' needs and demands.

Textiles with pre-woven starting borders are often regarded as evidence of the use of a warp-weighted loom. Some of these borders are made by card weaving (a common technique for narrow strips and belts). But there are other types of starting border, and one should not make an *a priori* assumption of a connection between card weaving and the use of the warp-weighted loom. As Ciszuk and Hammarlund stated, "All types of selvedges seem to have been used in connection with the warp-weighted loom but this use seems thus to be associated to the regional craft tradition rather than the type of loom."[19] Finds of rows of loom weights indicate that the warp-weighted looms could reach up to several meters in width, suggesting that more than one weaver would have worked together.[20] According to early medieval Nordic texts, a width of 2 to 3 aln (about 1–1.5 m) was appropriate for weaving on this type of loom.[21]

Pictorial evidence for the warp-weighted loom appears during the Late Bronze Age in the form of rock carvings in northern Italy, dated to the 14th century BCE. This loom may also be represented in a Cretan Linear A sign.[22] Later, the warp-weighted loom is depicted on Greek vases and is illustrated in engravings from 18th-century Iceland. Despite the differences of the depictions of warp-weighted looms in various time periods and regions, there are also interesting similarities among them. For example, all the loom weights are generally shown as being of the same size and hanging side by side.[23]

20.5 Other Looms That Use a Weighted Warp

There are other kinds of looms that use weight to create tension on the warp, such as relatively recent looms in Egyptian weaving workshops.[24] The loom is a type of pit-loom (section 14.3), in which the warp is kept taut by a few very heavy loom weights, often just one or two. The warp threads are separated into different sheds on heddle rods that are operated by a treadle system. This type of loom can be operated by a single weaver, which represents a labor-saving loom compared to the warp-weighted loom discussed above.

20.6 Frame Treadle Looms

Scholars generally assume that a frame treadle loom was first introduced into western Europe by the beginning of the 11th century CE at the latest. The exact route by which it arrived in Europe is unknown, but it is possible that this type of loom was already used in earlier time periods in the eastern part of Europe.[25] Over time, this newer loom gradually replaced the warp-weighted loom in all parts of Europe. It became the standard loom both in domestic and workshop contexts until relatively recent times. It later disappeared when the Industrial Revolution, which brought about the commercialization of weaving, led to the near-extinction of domestic weaving in most parts of Europe.

Notes

1 Hoffmann 1964.
2 Broudy 1979.
3 Broudy 1979, 38; Barber 1992, 83.
4 Barber 1992, 84.
5 Vogelsang-Eastwood 1992, 28–29.
6 Barber 1992; Vogelsang-Eastwood 1992.
7 Barber 1992, 84.
8 Bani Hamida weaver, personal communication to the author, 29 March, 2014. See also http://www.traditionaltextilecraft.dk/386325174.
9 Boyce, Kemp, and Vogelsang-Eastwood, 2001; Ciszuk and Hammarlund 2008.
10 Broudy 1979, 44; Barber 1992, 113.
11 Wild 1970, 68.
12 Mannering 2011.
13 e.g. http://www.georgeweil.com/ProductDetail.aspx?Menu=1&Level1=89&Level2=1256&Level3=0&PID=5191 (access April 2017).
14 Mokdad 2014, 19.
15 Dokumacı Fadime Koyuncu, communication to the author, Çavdar, Turkey, 2015.
16 Cutler 2012.
17 Andersson Strand and Heller 2017.
18 Barber 1992.
19 Ciszuk and Hammarlund 2008.
20 e.g. Belanová-Štolcová and Grömer 2010.
21 Hoffmann 1964; Østergård 2003.
22 Del Freo, Nosch and Rougemont 2010.
23 e.g. Hoffmann 1964; Barber 1992, 81–115; Gleba 2008, 29–33.
24 e.g. Broudy 1979.
25 Hoffmann 1964; Clarke 1986, 132; Coatsworth and Owen-Crocker 2017, 20.

21. Voice of the Weaver: Two Ancient Tales

Tibullus, *Elegies* 2.1, 1st century BCE
Translation from Latin by Magdalena Öhrman

In a poem celebrating the simple life of the Roman countryside, the poet Tibullus describes the typical tasks (and perhaps also enjoyments) of a woman responsible for cloth production. As the original Latin texts reflect the soundscape of weaving with the clanking of moving heddle bars and light clattering of loom weights, the poem shows both in form and content that weaving, weavers, and looms were self-evident parts of daily life in the Greek and the Roman worlds as depicted by artists on contemporaneous ceramics (figs. 21.1–21.3).

… here we also find the women's work,
the daily allotment of wool and the distaff,
here also the weaver,
who sings as she busies herself with constant craft,
while the loom resounds with the tinkling clay loom weights.

Fig. 21.1 Circe's loom. Kabeiric scyphus with Odysseus and Circe, 4th century BCE. (afer Broudy 1979, plate 23).

Darraðarljóð (Spear Songs), _Njal's Saga,_ chapter 156, 13th century CE
Translation by George W. DaSent (1861)

On Good Friday an event happened in Caithness that a man whose name was Daurrud went out.

He saw folk riding twelve together to a bower, and there they were all lost to his sight. He went to that bower and looked in through a window slit that was in it and saw that there were women inside, and they sang songs, and he learnt them by heart.

See! warp is stretched
For warriors' fall,
Lo! weft in loom
Tis wet with blood;
Now fight foreboding,
'Neath friends' swift fingers,
Our gray woof waxeth
With war's alarms,
Our warp bloodred,
Our weft corseblue.

This woof is y-woven
With entrails of men,
This warp is hard weighted
With heads of the slain,
Spears blood-besprinkled
For spindles we use,
Our loom ironbound,
And arrows our reels;
With swords for our shuttles
This war-woof we work;
So weave we, weird sisters,
Our war winning woof.[1]

Then they plucked down the woof and tore it asunder, and each kept what she had hold of.

Notes

1 "Woof" is an archaic term for weft, but in the context of this poem it refers to the woven cloth.

Magdalena Öhrman (PhD, Lund University) is a senior lecturer in Classics at the University of Wales Trinity Saint David (UK). In 2016–2018, she was a Marie Sklodowska Curie research fellow at the Centre for Textile Research in Copenhagen, where her project Textile Reflections examined multi-sensory representations of textile crafts in Roman literary sources (fig. 21.2).

The special thing about this [warp-weighted] loom is that it is the loom weights that make the magic happen—everything that you need to do, however complex a weave you are working on. Warp-weighted looms are very good for weaving twills because the sheds open extremely wide. And you work standing up weaving upwards. This is the other way around from a lot of vertical looms otherwise.

Fig. 21.2 Eva Andersson Strand and Magdalena Öhrman (right) preparing the warp-weighted loom for the exhibition *A World of Looms*, 2018 (photo: Magdalena Öhrman).

I am not a professional weaver, but learned to weave to better understand ancient texts about weaving, as in the Graeco-Roman world, this was very much part of everyday life. My colleague Eva [Andersson Strand] learned how to weave on the warp-weighted loom to better understand the archaeological evidence for weaving.

The warp-weighted loom is very versatile. For someone with basic experience of working on a treadle loom, it is really striking how hands-on one can be with the warp in this loom. Compared to looms where the warp is spanned between two bars, you have more options if something goes wrong. If the warp is not working the way you intended, it is easier to correct tension by re-adjusting the loom weights.

Fig. 21.3 Telemachus and Penelope at the loom. Artist's interpretation of a painting on a Greek drinking vessel, c. 460 BCE–450 BCE, in the Chiusi Museum, Italy (afer Broudy 1979, plate 24).

The Americas

22. The Andean Loom and the Making of Four-Selvedged Cloth
    Elena Phipps

23. Voice of the Weaver: Cusco
    Flora Callanaupa, Yanet Soto

22. The Andean Loom and the Making of Four-Selvedged Cloth

Elena Phipps

Weaving traditions in the region of the Andes are among the most diverse in the world, involving simple and complex weaves, recognized for their beauty and technical achievements (fig. 22.1). The region, whose history spanned millennia of cultural activities that extended thousands of kilometers, witnessed the development of state-level economies and empires. Throughout this time, textiles served an essential role in the cultures, as part of the life of the living and the dead, forming the basis of economic, political, social, and artistic constructions.

Strong regional traditions developed within the vastly diverse ecological zones—from dry coastal deserts to highland peaks and valleys. The native cotton, *Gossypium barbadense*—a long stapled fiber that grew naturally in a number of different hues, was a primary fiber used on the coast. The fine hairs of the various camelids—llamas, alpacas, vicuna, and guanacos—which thrived in the upper altitudes, formed the basis for the weaving traditions of the highlands (fig. 22.2).

Fig. 22.1 Mantle with discontinuous warp and weft technique and embroidered borders, Paracas Necropolis (Wari Kayan), south coast of Peru, 0–100. Collection of the Museo National de Arqueología, Anthropología y Historia, Lima, RT3554 (after MNAAH/Daniel Giannoni, 2008).

Facing page: *Tocapu* tunic. Peru, Inca period, 1450-1540. Tapestry weave, camelid hair and cotton 90.2 cm x 77.2 cm. Collection of the Dumbarton Oaks. P.C. B. 518 (photo: Dumbarton Oaks).

Fig. 22.2 Camelids, alpaca (photo: J. Blassi, after Flores Ochoa, MacQuarrie, and Portús 1994).

Evidence of the development of these textile traditions comes from preserved archaeological fragments as early as 4500 BCE when complex constructions of looping with wrapped elements and weft-twining with diverted warps composed textiles of religious significance that mirrored the images of stone-carved deities.[1] These processes developed into simple and complex weaves that enabled the weavers to create images of power and religious significance formed within the structure of the cloth.[2] Early in this development, compound weaves of double cloths, triple and quadruple cloths, and even textiles with five sets of warps and wefts (fig. 22.3) were produced, as well as supplementary and gauze weaves. Additional complex weaves include those with complementary warps, with warp substitution and warp-faced tubular double cloths (sometimes with six sets of warps), among many others, most of which developed prior to 300, and all, for the most part, composed as uncut, four-selvedged cloths.[3]

22.1 The Loom

Ed Franquemont, a scholar who worked in the field of highland Peru for many years, remarked that an Andean loom is only a loom when it has a warp on it; otherwise, it is just a series of sticks (fig. 22.4).[4] His work and that of several others have shown that the intensive complex work of Andean weavers for the most part is not the result of the loom, but rather, comes from the conceptual ability of the weaver, through the use of systems of logic, mathematical counting, an understanding of symmetries and three-dimensional pathways, and the acute manual dexterity of the artisan that is achieved through a sequential and cumulative learning process over time.

22. The Andean Loom and the Making of Four-Selvedged Cloth

Fig. 22.3 Overall and detail: band, camelid hair, five color quintuple-cloths (six sets of warps, five sets of wefts), Peru, Nasca 200–400. Collection of the Fowler Museum at UCLA, X65-15408 (photo: Fowler Museum).

Fig. 22.4 Photo titled "Laying out three loom bars for special warp (Q'eros) mid-1950s" (after Cohen 2010).

22.2 Andean Looms

Fig. 22.5 Detail of weaving scene, fine-line painting on a pottery vessel, Peru, Moche Culture, 2nd–8th centuries. Trujillo, Peru. British Museum, London, Am1913,1025.1 (photo: British Museum).

Andean looms (cat. 49) can simply be composed with a minimum of two sticks, and can be used horizontally staked to the ground, standing vertically upright or leaning against a wall or tied to the weaver's waist. The appropriate tension is achieved in several ways according to the type of loom, the weaver's body, and the selection of materials.

Information about the types of looms used by ancient Andean weavers comes from archaeological records of preserved textiles, loom components, and representations of looms and weavers in other media such as pottery or metalwork. A weaving workshop is depicted on a slip-painted 2nd–8th-century ceramic vessel from the north coastal Moche Culture of Peru (fig. 22.5).[5]

The weaver is depicted seated on the ground, with a backstrap loom with one end tied around her waist, and the other attached to a fixed point in the ceiling post. The textile in process of weaving is patterned, and there is some kind of representation of a pattern model that also seems to be depicted. While few extant textiles have been found from this period, some fragments of the types of cloth these weavers appear to be weaving are known, generally in tapestry or supplementary weft-patterned weaving.[6]

The backstrap (or body-tensioned) loom is ubiquitous throughout the Andes, and continues to be used in both the highlands and the coast up to the present day. For these looms there is one fixed end of the warp, tied to tree or post of some type, with the other end attached to a belt around the weaver's back. The weaver adjusts her body position to increase or decrease tension in the warp, by leaning forward or back, facilitating the lifting of heddles or shed rods, or individual patterning. This type of backstrap loom was used for a variety of weave types, including plain weaves of weft or warp orientation, supplementary weft weaves, double cloths, and gauze weaves, among others.

In the 12th–14th centuries, there is evidence of another type of loom—an X-shaped frame loom—used on the north coast, seen in the miniature silver Chimu burial offering, that also included many of the tools, such as weaving combs, weft bobbins, weaving sword, among others. While this type of loom is less known from actual preserved looms, it would have held the warp at a fixed tension, unlike the variable tension of the backstrap loom. Because of this fixed tension, it may have been used for tapestry weaving.[7]

In the highlands, highly refined tapestry production was developed to produce densely woven and aesthetically sophisticated garments used by the ruling Wari and Tiwanaku Cultures of the 7th–9th centuries. While no looms have been preserved, the creation of a textile of this quality, with hundreds of yarns per inch of packed wefts, probably required a fixed tension frame-type loom. Woven in two long and narrow symmetrical lengths which would be stitched up the middle to form the garment, the loom must have been very wide, and short. These garments have distinctive warp selvedges—the lower part is uncut, while the upper selvedge was cut from the loom, but the warp yarns re-entered in diagonal to form a finished edge.[8]

The Inca, who hundreds of years later linked their mythological origins to the highland Tiwanaku Culture also practiced this meticulously controlled tapestry weaving, creating garments that were double-sided and completely finished inside and out (see *tocapu* tunic on the facing page). Great efforts were put into the production of very fine textiles to be used by the nobility of the Inca empire. Royal workshops which likely housed hundreds of weavers have been excavated. The densely woven tapestry garments, *cumbi*, made by the "chosen women" of the Inca, are unparalleled in their quality, with sometimes over 250 yarns per inch.[9] To weave this type of textile a very strong fixed tension must be maintained in the warp. The only record of a possible Inca tapestry loom comes from a ceramic vessel, which depicts two weavers—one on each side of the loom.[10]

From the early colonial writings of the 16th century, looms and weavers were the subject of the documentation of the Andean world by Spanish friars and indigenous scholars. Both backstrap and upright looms are depicted by Guaman Poma de Ayala in a 1615 manuscript written for the Spanish king (fig. 22.6).[11] The Spanish commissioned former Inca royal weavers to produce large wall-hanging tapestries in the European style, produced no doubt on these very large upright looms.

The warp is generally not rolled in this type of loom, rather the weaver shifts his position to reach the top of the weaving. This style of weaving, to produce rugs and tapestries continued until recent times, persisting in some areas of Bolivia for example (fig. 22.7).

Though weft-faced tapestry was produced in the highlands, more pervasive weaving traditions involved the production of warp-faced and warp-patterned weaving. We know that today highland weavers use at least three types of looms—the backstrap loom, the upright frame loom, and the horizontal staked-out ground loom (fig. 22.8). The four-stake ground loom provides a fixed tension to the densely set warps required for the warp-faced weaving. The warp bars are attached to the stakes, and the weaver works seated on the ground. It is often used for warp-patterning, which requires the lifting of selected warps for complementary-warp patterning techniques. Some flexibility and control

Fig. 22.6 Weaver seated at upright tapestry loom. Guaman Poma de Ayala, 1615 manuscript. Royal Library, Copenhagen, Gl. kgl. S. 2232, 4° (photo: Royal Library, Copenhagen).

171

Fig. 22.7 Tapestry weaver, Tarata, Bolivia, 1998 (photo: Elena Phipps).

Fig. 22.8 Aymara weaver with four-stake ground loom, 1983 (photo: Amy Oakland).

Fig. 22.9 Left: cotton textile in process of weaving, including loom bars, heddle rods, shuttle sticks. Unknown place of origin and date, possibly south coast of Peru. Right: detail view of loom cord and warp selvedge. Collection of the Fowler Museum at UCLA, X.65.14334A (photo: Fowler Museum).

over the tension is required, and the use of overtwisted camelid hair yarns, which in themselves have some "give," facilitates the weaving process.

While distinctive weaving traditions developed between the coast—with its characteristic weft-orientation in its methods—and the highlands—with its warp-faced weaving, one shared feature that characterized the entire region was the making of four-selvedged cloth, used complete as it is taken from the loom. Using this method, each textile that was woven was created for a specific purpose, reflected in its size, format, and design. Cloth was never cut: garments formed from webs of cloth may be stitched but were not tailored. The technique requires specific focus and attention to the preparation of the warp, and its positioning on the loom (fig. 22.9). Warps are prepared for size and format their intended final product. Prior to warping, a prayer and an offering of alcohol or coca leaves are often made.

Rather than winding the warp around the loom bar, a heavy heading cord is used to hold the uncut warp ends to the bar, which is then secured to the loom. In this way the weaver is able to weave from one end of the warp to the other. Then, without cutting, the web can be removed from the loom after weaving, intact. As the warp is not cut, weavers prepare heddles anew for each new warp on a loom.

Fig. 22.10 Photo titled "Inserting the final passes of weft using a needle, near Ocongate, 1956" (after Cohen 2010).

Generally, a shed stick and one set of heddles is sufficient for plain weave, though other types of weaving may utilize more. For warp-patterned weaving with multiple sets of warps, there may be further devices used to help maintain the order of the color sets. Most complementary weaving utilizes paired colors, and laying out the warp sets and color sets are a first step to the weaving process. This is extremely important as the weaver uses her hands and fingers, and perhaps a small bone tool, called a *wincha*, to lift the correct yarns to create the designs. This tool has been used for thousands of years and continues to be used to the present day.

In most cases, the weaver begins at one end of the warp, weaving a short length of web, then turns the loom around and begins weaving from the other end. A needle is required to insert the final wefts at the point where the two leading edges meet, because the sheds in the unwoven part of the warp become too narrow to insert a shuttle (fig. 22.10). These "terminal areas" are especially visible in warp-patterned textiles, as the weaver is unable to lift the threads for patterns in the remaining unwoven section of warp. In garments, such as a woman's mantle, the two four-selvedged panels are seamed together, placed so that the terminal areas are countered to each other.[12]

A woman's mantle, called a *lliclla* in Quechua or *phullo* in Aymara—the two main languages of the Andean region—is composed of two units of cloths: each is called a *callu* meaning half of something. Even though each is woven as a complete cloth—with all four edges intact—it is conceived as incomplete without its pair. Designs on these cloths are asymmetrical when viewed in its single units: once seamed down the center, the design gains symmetry, and the garment gains its equilibrium.

Several looms have been preserved which show a special type of weaving that has been practiced since at least 300 BCE that uses a series of horizontal sticks which function as a scaffold, enabling the weaver to prepare the color changes of the warp, in sections.[13] Sometimes made for ritual cloths, the tradition has been revived in recent years in the Cuzco region (cat. 49).

22.3 Looms and Weaving Knowledge

The people of the Andes had no writing until its introduction by the Spanish in the 16th century. The development of complex textile systems, therefore, relied on strong traditions of learning and knowledge, passed down through the generations. Some of these systems involved conceptualizing design sequences, symmetries and dualities in their making, often from memory, while creating complex hand-picked patterning. These memory systems, learned from a young age, developed in stages, and became part of what Desrosiers termed the "logic of weaving."[14] We know that these systems were in use in the 1590s, when Martin de Murúa, an Augustinian friar recorded the "memory" for the making of a belt to be worn by the Inca queen for the annual Corn Festival.[15] While many cultures around the world utilize complex mechanisms on the loom for creating patterning systems in the Andes, it is the weaver herself who holds that knowledge.

Notes

1 Conklin 1978; Doyon-Bernard 1990.

2 Bird 1963.

3 "Every piece of cloth they made, for whatever purpose, was made with four selvedges. Cloth was never woven longer that what was needed for a single blanket or tunic. Each garment was not cut, but made as a piece as the cloth came from the loom and before weaving it they fixed its approximate breadth and length" (Garcilaso de la Vega 1609, book 4, chapter 13, p. 214).

4 Dransart 2007, 174.

5 British Museum, London. Acc no: Am1913,1025.1.

6 Conklin 1979; Donnan and Donnan 1979.

7 Miniature silver weaving tools and X-shaped loom. Chimu (North Coast) 1150–1450. Collection of the Peabody Harvard Museum, Massachusetts. Acc no: 48-37-30/7162.

8 See Bird and Skinner 1974.

9 See Phipps 2004, catalog number 3.1, pp. 130–131.

10 Vanstan 1979.

11 Guaman Poma de Ayala 1615/1616 (http://www.kb.dk/permalink/2006/poma/info/en/frontpage.htm).

12 Phipps 2013

13 See Strelow 1996, fig. 1a, p. 10. See also Phipps 1982.

14 Desrosiers 1997.

15 Desrosiers 1986.

23. Voice of the Weaver: Cusco

Flora Callanaupa, age 53, comes from the Chinchero, Cusco in Peru. She was born in a well-known place for textiles and weavers. She learned weaving and developed her appreciation of her weaving culture starting from seven or eight years old. As a young girl, she already identified herself as a weaver. She is now one of the weavers at the Centro de Textiles Tradicionales, Cusco, Peru (fig. 23.1).

I learned how to weave when I was 7 or 8 years old with my mom. I first learned to weave very small pieces, it was like playing. I feel very well and focused when I weave. All my imagination comes into play when I weave and I feel motivated to achieve my goals. The most important thing about weaving for me is to be able to share my technique and culture with others to keep the tradition alive. My market is mostly north American visitors, but also other foreign visitors. My dream is to better our textiles and continue to help communities of weavers around Peru. Also, to travel around the world to spread the knowledge about our ancient textile tradition. [At the China National Silk Museum] it was very interesting to see and learn about a completely different culture, a new world of textiles and weavers with such fine and complex textiles. It was amazing to see that the world is full of such different textiles.

Fig. 23.1 Flora Callanaupa during a weaving demonstration at NSM (photo: Elena Phipps).

Yanet Soto is a weaver. She works at the Centro de Textiles Tradicionales, Cusco, Peru (fig. 23.2).

The most important for me is the traditional weaving and the techniques in each town.

Fig. 23.2 From left: Elena Phipps, Flora Callanaupa, and Yanet Soto (photo: Elena Phipps).

The Jacquard Loom and After

24. The Development of the Jacquard Loom
   Guy Scherrer

24. The Development of the Jacquard Loom

Guy Scherrer

A "figured" or "patterned" textile has a pattern that is constructed of complex yarn crossings. Preparing the design of such a textile involves exact planning, for example by using a graph paper plan where each square represents a crossing between a warp yarn and a weft yarn. While some weavers can memorize the design plan for small patterns, or copy an older textile, in many cases the design is stored on the loom using one of several alternative systems multiple shafts (heddles linked to treadles), drawloom systems using cords or rods to store designs, punched cards (Jacquard looms) or a computer (digital looms).

24.1 Before the Jacquard Loom

In Europe, until the beginning of the 19th century, three types of looms were used to weave figured fabrics: the multi-shaft loom, the button drawloom, and the drawloom. The first two types were used to weave less complex figured fabrics and the weaver can work alone. On the multi-shaft loom, the warps are lifted using many different shafts. Up to around 80 shafts are possible, limited by the space such shafts take up and the complexity of working many treadles. On the button drawloom, all the necking cords lifting up for the first weft shot are attached to the button number one, second weft shot to the button number two, and so on. The drawloom proper (fig. 24.1), was designed for weaving the most complex patterns. The pattern is stored by means of a cords system[1] that is arranged on the right or left of the loom. At least one drawperson, in addition to the

Fig. 24.1 Drawloom. Maison des Canuts, Lyon (photo: Guy Scherrer).

Facing page: polychrome silk with floral design on blue ground. Supplementary-weft patterning, France, 1760s.

181

weaver, is required to operate the pattern cords. Pulling a pattern cord lifts a group of warp yarns. This activity is physically demanding, making it hard to find people who were willing to take a job as a drawperson in 18th-century France.

It took almost a century of research to find a replacement for the drawperson, attempts that eventually led to the innovation of a card-reading machine. The first device of this kind was developed by Basile Bouchon, a Frenchman from Lyon, in 1725. A second was made by Jean Baptiste Falcon, also from Lyon, around 1735. Another important loom was made by Jacques de Vaucanson in Paris in 1748. These looms are some of the most prominent ones; there are many other engineers whose names are known because of awards they received for their work, but the details of their mechanisms are not clear.

24.2 The Jacquard loom

A Jacquard loom (fig. 24.2, cat. 50) is a loom controlled by a card-reading device, called the Jacquard machine. This device was developed in Lyon between 1806 and 1817 by several engineers, but because of its complexity it did not run efficiently until 1817. While the Jacquard machine received its name from Joseph Marie Jacquard, it was the engineer Jean Antoine Breton who in fact played a more important role in its development and registered some key patents between 1815 and 1817. The association of Jacquard's name with this invention is largely a matter of chance.

The Jacquard machine employs a binary code system using cards that are punched with holes set at certain distances. The presence of a hole (=1) codes for the lifting of a necking cord,

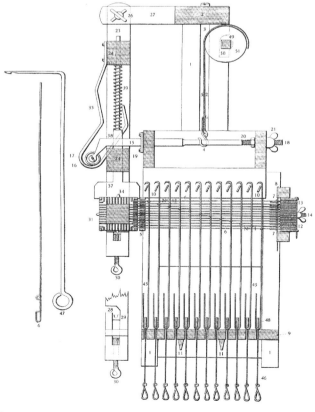

Fig. 24.2 Left: Jacquard loom. Maison des Canuts, Lyon (photo: Guy Scherrer). Right: illustration of Jacquard machine (after Loir 1926, 8, plate II).

connected to one or more warps, and an absence of hole (=0) indicates that the necking cord stays in place. In term of the mechanics, the holes on the cards control horizontal needles, and the movements of these needles control vertical hooks that are attached to necking cords and heddles. If there is a hole, a needle goes through it, and the hook remains stationary. In this position, the hook is caught by a rack system and lifted up. Consequently, the corresponding necking cord, heddles, and yarn are raised. If the needle hits a card face with no hole, the card pushes the needle, and the needle pushes the hook. As a result, the rack does not catch the dislocated hook, and the heddle remains in place.

One hook can lift one necking cord, or it can lift several if the loom is set up with several comber units, for instance to weave repeats of the design along the weft direction. One card is used for each pass of the shuttle, so the number of cards corresponds to the length and the fineness of the pattern repeats. The cards are made of cardboard; they are kept in the correct sequence for a particular pattern, concertina-fashion, by being bound with laces.

The original Jacquard machine was made of wood, and the needles, hooks and some pieces were made of iron. The number of hooks could be up to 1,200. For a 1,200-hooks machine, the cards were 80 cm long by 8.7 cm wide, with a card thickness of 1 mm and a hole diameter of 0.45 cm, and a distance between the centers of two holes of 0.675 cm. The Jacquard machine spread very quickly—first in Europe, then throughout the world—partly because of its usefulness in commercial weaving, and partly because Jacquard did not patent the original machine, with Breton patenting only some of parts of it.

24.3 Modifications to the Jacquard Machine (19th Century)

The dissemination of the Jacquard machine spurred many new patents on the individual components that were meant to improve the performance of the machine. Many patents centered on efforts to replace the cardboard cards, since the large cards were costly to produce.

Before 1820, loom manufacturers in Germany and Austria produced wooden Jacquard machines. The advantages of wood were resistance to rust and low price. These wooden looms were suitable for low-intensity use, for example in provincial weaving workshops. Around 1835, fully metal machines appeared. They were more stable to temperature and humidity fluctuations, more durable, and, above all, more precise, allowing the use of smaller cards. The Vincenzi machine was patented in 1869, using smaller cards than the Jacquard machine. For a 1320-hooks machine the card system was 37 cm long by 7 cm wide. The hole diameter was 0.3 cm and the distance between the centers of two holes was 0.4 cm (fig. 24.3).

In 1843, a patent was registered to allow the hooks in the Jacquard machine to "raise and lower," instead of to "raise and stay in place." The ability to lower the hook is better for mechanical looms,

Fig. 24.3 Comparison of pattern cards.
Top: Jacquard card with 612 hooks.
Middle: Vincenzi card with 1,320 hooks.
Bottom: several Verdol papers with 1,344 hooks.
(photo: Guy Scherrer).

because the warp tension remains equal between the "up" and "down" yarns. Another improvement was patented 1854. Until this time, a hook used twice in a row had to be raised twice: first to lift it up, then to lower it, and then to lift it again. With the improved loom the hook remained in the "up" position between these two uses, increasing the weaving speed.

Also in 1854, Bonelli patented an electric machine to replace cards. In his machine the hooks were controlled by electromagnets; the design was recorded on a cylinder rather than a card, with electrical insulating or conducting parts denoting the design. Though this version was unsuccessful, it can be considered as an ancestor of the "digital Jacquard" machine. It is also important as an early example of an electric storage medium, similar in basic principle to magnetic storage media that are used today in computing.

Attempts to modify the cards by using paper, which is cheaper and lighter than cardboard, were first patented by Skola as early as 1819. Similar patents followed until 1883 but none proved commercially viable. The main challenge was in creating a stable-enough paper to withstand variations in humidity. This proved impossible even with coatings and varnishes.

24.4 The Verdol Machine (1883)

Jules Verdol, an engineer in Paris, was one of the inventors who tried to replace cardboard cards with paper cards. Having worked on this problem unsuccessfully since 1860, he concluded that it was necessary to design a new type machine instead of adapting the Jacquard one. He later established his own company and, in 1883, registered a patent for a new metal machine that read a continuous paper roll instead of cards bound with laces. The Verdol machine (fig. 24.4) further miniaturized the program system: each weft shot was controlled by 1,344 hooks; the continuous paper roll was 47 cm long and 2.7 cm wide. The hole diameter was 0.2 cm and the distance between two holes, center to center was 0.3 cm.

The efficiency and compactness of the Verdol machine were advertised in the company's 1886 catalog. This states that for a 1,200–1,300 hook machine and 1,000 cards (=1,000 shots, 1,000 wefts), the card weights are: 53 kg for a Jacquard machine, 15 kg for a Vincenzi machine, and a mere 2 kg for the Verdol machine. The Verdol machine was three times more expensive to buy than a wooden Jacquard loom, but it required lower maintenance. The cost difference could be covered after making 12,000 cards, representing around ten different designs. In addition to the

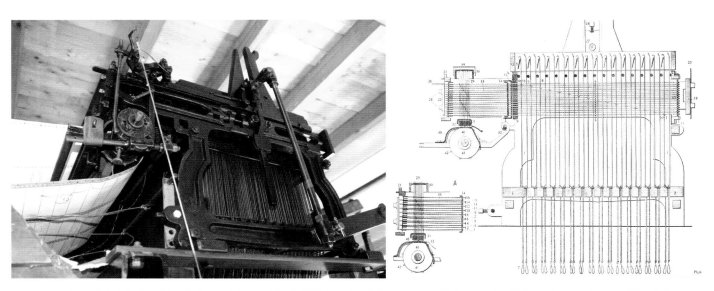

Fig. 24.4 Left: Verdol machine with 1,344 hooks (photo: Guy Scherrer). Right: illustration of Verdol machine (after Loir 1926, 15, plate IV).

Verdol machine, the Verdol company also sold the punching equipment for making the program rolls and, of course, the paper.

At the beginning of the 20th century, the company moved from Paris to Lyon, where the heart of the European silk weaving industry is located, and opened branches in textile centers in Barcelona (Spain), Como (Italy), Elberfeld (Germany), Kyoto (Japan), Mahr Schonberg (Austria), Moscow (Russia), Paterson (the United States) and Zürich (Switzerland). Some branches had their own punching machines and provided a service for weavers by reading their graph-paper plans and punching program rolls.

Within a century, from 1883 to 1983, the Verdol company built nearly 50,000 of these machines, an astonishing total. Between the first machine and the last one, many improvements were made, including a faster speed, ranging up to 600 rpm for label-weaving looms with 448 hooks) and 410 rpm for looms with 2,700 hooks. Remarkably, the Verdol paper roll standard has not changed, and it is still in use today worldwide. When weavers say "Jacquard cards" and "Jacquard looms" what they are actually referring to are "Verdol paper rolls" and "looms controlled by Verdol machines." The original wooden Jacquard machines and the Vincenzi ones are no longer used, except in rare cases and as museum demonstrations. In 1983, Stäubli bought the Verdol company, later becoming known as Stäubli-Verdol. The production of Verdol paper machines continued until 1997. From its inception in 1883 until 1914, the company received many awards at international exhibitions.

24.5 The "Digital Jacquard" Machine

From the end of the 20th century until the present day, loom speeds have increased from 300–500 rpm to 1,000–1,200 rpm, driven by demand for faster production and lower unit cost. Because of the high speed, weaving figured fabrics using the Verdol machines with punched papers soon reached the limits of what paper rolls are capable of. The first "digital Jacquard" machine appeared in 1979, first for making ribbons, with a version appearing in 1987 for wider fabrics. In fact the term "digital Jacquard" hardly does justice to the capabilities of these looms.

As mentioned, Jacquard systems are based on a binary code (the presence of a hole triggers the lifting of the yarn, while the lack of a hole avoids the action). It is easy to imagine the transition of this coding to computers. Jacquard machines today no longer read punched cards or paper, but rely on magnetic or optical storage. These machines comprise stationary racks and hooks. The computer controls electromagnets and these electromagnets in turn control the hooks. It is a modular design, so if a modular unit of eight hooks fails, it is easy to swap it out and restart production. This kind of machine also uses very little energy to raise the hooks and can do so extremely fast. Electronic Jacquard machines may have up to 26,000 hooks and run at speeds between 600 and 1,000 rpm.

In the past, Jacquard machines had limited numbers of hooks (1,200–1,300) controlling 8,000–12,000 warp yarns, using different harnesses and pattern cords according to the design symmetries (for example straight or point repeats). Different fabrics would require different harnesses and cord arrangements. Today with a very large number of hooks, more precise and flexible control of warps is possible, so the same basic harness remains on the loom and many kinds of design can be made with it. Changing a design simply entails changing a file in the computer.[2]

The most recent design evolution is the fully electronic Jacquard machine (cat. 51). With this machine, it is possible to control every single warp individually. These machines do not use racks and hooks. Instead they have small brushless actuators (motors) with one motor per yarn. These allow each individual yarn to be lifted or depressed to any desired height (variable shed), allowing the possibility to weave 3D fabrics such as tubes and airbags.

Notes

1 In CIETA terminology, these are necking cords, pulley cords, simple cords, and lashes.

2 For further reading, see Diderot and d'Alembert 1765; Adrosko 1982; Charlin 2003; Scherrer 1993, 2006, 2012.

Catalog

East Asia: Archaeological Evidence from China

East Asia: Present-Day Looms from China

East Asia: The Korean Peninsula

Mainland Southeast Asia

Insular Southeast Asia

South Asia

Central and Southwest Asia

Africa

Europe

The Americas

The Jacquard Loom and After

List of Looms in the Catalog and Their Locations

1 Tianluoshan Loom
2 Tianluoshan Ivory Weft Beater
3 Liangzhu Jade Loom
4 Jing'an Wooden Loom
5 Shizhaishan Bronze Loom
6 Luobowan Wooden Loom
7 Laoguanshan Pattern Looms
8 Han Dynasty Oblique Treadle Loom
9 Han Dynasty Vertical Treadle Loom
10 Backstrap Treadle Loom
11 Balanced Paired-Treadle Loom
12 *Kesi* Loom
13 *Dingqiao* Loom
14 Taiwan Atayal Loom
15 Dong (Kam) Brocade Loom
16 Dai Jinghong Brocade Loom
17 Zhuang Binyang Bamboo Cage Loom
18 Zhuang Jingxi Brocade Loom
19 Velvet Loom
20 Damask Drawloom
21 Gauze Drawloom
22 Lampas Drawloom
23 *Yun* Brocade Drawloom
24 Korean *Bettle* Backstrap Treadle Loom
25 Tai Frame Loom
26 Lao Tai Frame Loom with Vertical Pattern Heddle

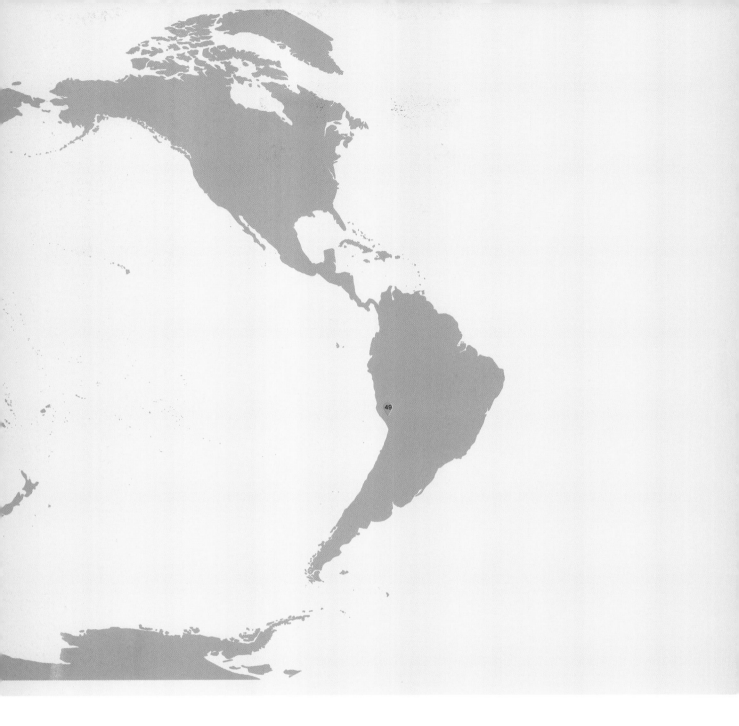

27 Tai Phuan Frame Loom
28 Khmer Frame Loom
29 Sulawesi Body-Tensioned Loom for *Tenun Mamasa*
30 Bali Body-Tensioned Loom for *Geringsing* Double Ikat
31 Sumba Body-Tensioned Loom for *Hinggi Kombu*
32 Simalungun Batak Body-Tensioned Loom for *Bulang* Textile
33 Palembang *Songket* Loom
34 Bengal Pit Loom
35 Gujarat Multiple Heddle and Treadle Loom
36 Varanasi *Jaala* Drawloom
37 Ancient Egyptian Ground Loom
38 Bedouin Ground Loom
39 Egyptian Carpet Loom

40 *Zilu* Loom
41 Margilan Ikat Loom
42 Ghana and Togo Loom (*Agbati*)
43 Central Ghana Loom (*Nsadua*)
44 Madagascan Loom for *Akotifahana* Textile
45 Madagascan Loom for *Akotso* Textile
46 Madagascan Loom for Raphia Textile
47 Ancient Warp-Weighted Loom
48 Aubusson Tapestry Loom
49 Chinchero Body-Tensioned Looms
50 Jacquard Loom
51 Computer-Programmable Loom

East Asia: Archaeological Evidence from China

Introduction

The first part of this catalog concerns archaeological remains from the Chinese mainland. Favorable preservation conditions in a few key sites, coupled with a long-standing practice of burying the dead with useful items for the afterlife, including (in some instances) looms, have left important evidence of early Asian looms from the Neolithic Period onwards. These remains are important for understanding the development of loom technology in the entire Asia region, as well as China itself.

1 Tianluoshan Loom

The Tianluoshan site, located in Yuyao city, Zhejiang province, was a prosperous village belonging to the ancient Hemudu Culture as early as 6,500 years ago. The inhabitants of this Neolithic-Period site are known to have practiced rice cultivation, as well as gathering wild plants, and fishing. The state of preservation of archaeological remains at Tianluoshan is exceptional; wet conditions at the site have preserved wood and bone parts that normally decay. These remains provide a full picture of daily life, which seems to have included several different kinds of weaving, net-making, and basketry. A variety of spindle whorls of different sizes and weights attest to the production of yarns of various types.

The loom parts excavated from the Tianluoshan site seem to come from simple backstrap looms.

Wooden loom components, excavated at Tianluoshan. Collection of the Zhejiang Provincial Institute of Cultural Relics and Archaeology (photo: Sun Guoping).

Facing page: pottery loom model, Han Dynasty (206 BCE–220 CE); H. 30 cm; L. 25 cm; W. 17 cm. Collection of the Musée National des Arts Asiatiques-Guimet (MNAAG) or Musée Guimet, Paris. Bequest of Krishnâ Riboud, 2003, MA 12024 (photo: RMN-Grand Palais/Thierry Ollivier).

With this type of loom, one end of the warp would have been tied to a warp beam, while the other end would have been connected to a cloth beam, attached to the waist of the weaver. Several pointed tools have also been unearthed this site, including what appears to be a warp separator to open the natural shed and a weft beater or "sword," similar to present-day weaving tools used in Southeast Asia. It is possible that the counter sheds would have been opened using a heddle as in simple looms in Southwest China and Southeast Asia (sections 6.1, 9.1 and 10.1), but this is difficult to ascertain since there are also many finds of pointed pick-like tools that could have been used as shed openers. Other tools that were unearthed include rods resembling warp or cloth beams, pegs that might have been used to secure a warp beam, a notched rod that might be a warp spacer, and a tool with pointed, forked ends that might be a spool or a temple. These tools are discussed in more detail in section 4. (LB)

2 Tianluoshan Ivory Weft Beater

A beautiful knife-like object made of ivory (labeled "dagger" in the museum exhibit at Hemudu) was unearthed at Tianluoshan. It is short and blunt, with an oval profile and decorated handle, finely carved in the shape of a bird. Its shape and size resemble a small weaver's "swords" that are still prevalent all across Asia and used for weaving belts and straps (see illustration on the right, below). The fine carving and the ivory material of this weft beater suggest that weaving in the ancient Hemudu Culture (5200 BCE–4200 BCE) was associated with a degree of status. (CB)

Carved ivory object excavated at Tianluoshan, Hemudu Culture, possibly a weaver's sword from a small belt loom (photo: Wikimedia Commons).

Small belt loom used by contemporary Dong (Kam) weavers in the Dimen area of Guangxi Zhuang autonomous region. Loom parts from the top: backstrap, cloth beam, weaver's sword, heddle, shed stick, warp beam (photo: Christopher Buckley).

3 Liangzhu Jade Loom

Jade ornaments from a simple backstrap loom of the Liangzhu Culture. Unearthed from tomb 23 in Fanshan, Yuhang, Zhejiang province (the length of the longest component is about 4.5 cm).

This is the earliest, complete—or nearly complete—loom discovered in China to date. It was unearthed in 1986 from a high-status burial (tomb 23) in Fanshan, Yuhang, located north of Hangzhou. The tomb is dated to the Neolithic Period, to early part of the middle stage of the Liangzhu Culture, about 4,800–5,000 years ago.[1] In total, 6 jade components consisting of three pairs of finials were unearthed. These were originally greenish in color, but have turned beige due to long burial. The pairs were found spaced 35 cm apart in the tomb. It is assumed that originally there were wooden rods in between, connecting the pairs of jade finials. The analysis of the cross sections (thickness and width) of the jade ornaments suggests that they corresponded to three loom parts: a warp beam, a cloth beam and a warp opener/beater. It is likely that the loom also had a rod for retaining the natural shed (shed stick) as well. The warp beam is flat on one side and semi-circular on the other, suitable for fixing the warp yarn and bracing the beam with the feet, which were positioned on the flat side. The cloth beam has two interlocking parts for clamping the woven fabric, and knobs at the end for securing a backstrap, which has not been preserved. The warp opener/beater has a flattened profile and rounded ends, and was used for enlarging warp openings, as well as beating-in the weft. The warp opener/beater has blunt ends and so is unlikely to have been used for selecting individual warps. This implies that some other devices must have been used to select warps for the counter-shed, most probably a heddle.

This loom looks very similar to backstrap foot-braced looms that are still used today by minority peoples on the islands of Hainan and Taiwan (cat. 14). It is operated as follows: a warp is wound in a circular fashion and then placed over the warp beam and cloth beam, clamping the two halves

A World of Looms: Weaving Technology and Textile Arts

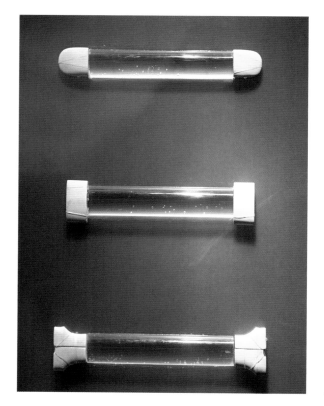

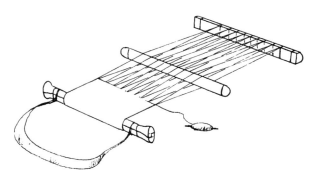

Above: the Liangzhu jade loom components arranged according to their original orientations, with perspex replicas of decayed wooden parts showing how they would have been used (top: L. 2.8 cm, W. 3 cm; middle: L. 4.5 cm, W. 2 cm; bottom: L. 4 cm, W. 3 cm). Collection of the Liangzhu Museum.

Below: reconstruction drawing of the Liangzhu loom.

of the cloth beam to hold it in place. The weaver then attaches the cloth beam to her waist using the backstrap. To open the natural shed she flexes her feet to tension the warp, inserting the warp opener/beater into the opening and turning it through 90 degrees to open the shed wider, then inserts one throw of weft, after which she uses the warp opener/beater to beat-in the weft. She then relaxes the tension in the warp and pulls up the heddle to open the counter-shed, removing the warp opener and placing it in this new opening, then turning through 90 degrees and inserting another throw of weft. After a short length of cloth has been woven, she loosens the cloth beam clamp and moves the warp around, continuing until the whole length has been woven. By this means she can weave cloth with a width not exceeding 35 cm, corresponding to the gap between the jade ornaments.

Before the Liangzhu loom was discovered, large numbers of probable loom parts had been found in sites such as Hemudu, but because the unearthed parts were not in complete sets and their uses are not obvious, it is difficult to make definitive restorations. The Liangzhu loom is the earliest group of loom parts that undoubtedly constitute a set, so it has very important scientific value for the research on weaving in the Liangzhu Culture and in the history of textiles generally.[2] The Liangzhu loom is the probable ancestor of many or all of the backstrap, body-tensioned looms that are used across East and Southeast Asia today. (LB)

4 Jing'an Wooden Loom

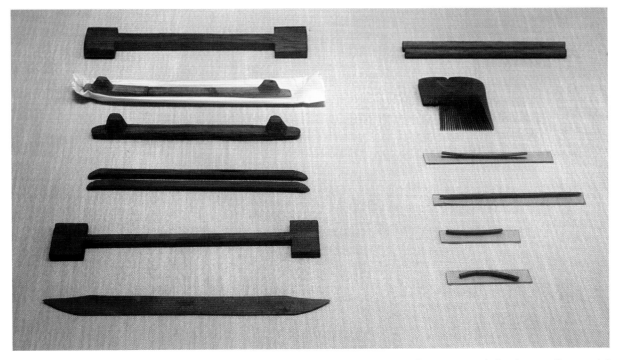

Wooden loom components, unearthed from Jing'an, Jiangxi province. Collection of the Jiangxi Provincial Institute of Cultural Relics and Archaeology.

At the end of December 2006, a large intact, sealed tomb mound from the Eastern Zhou Dynasty (770 BCE–256 BCE) was found in Jing'an, Jiangxi province. A year later, the Jiangxi Provincial Institute of Cultural Relics and Archaeology and the Jing'an Museum began the excavation at the site. The tomb yielded 47 coffins and a variety of precious objects made of gold, bamboo, lacquer, wood, bronze, jade, and porcelain. These items provide a unique picture of the culture of southern China during the Eastern Zhou Dynasty. Some of the most striking finds are the textiles and textile tools.

Nearly a hundred textile-making tools were found, indicative of the high importance of weaving during this period. Most artifacts were placed in bamboo containers at the feet of the deceased, but some were also found outside the coffins. The finds included wooden winding plates, shuttles, beaters, bamboo tubes and bobbins, as well as ceramic spindle whorls. All the loom parts are rather small and crudely made, a common feature of burial goods. The burial looms are not actual functional looms; instead they represent looms to be used in the afterlife. The loom components correspond to body-tensioned looms, as can be deduced from the shape of the cloth beams that have knobs for attaching backstraps. The number and variety of textile tools indicate that Jing'an may have been a textile weaving center and an important area for textile production during the Eastern Zhou Dynasty.

The loom parts included I-shaped warp beams with flat ends, resembling paddles. This feature tells us that the warp beams were designed not to rotate during weaving (the "paddles" preventing movement). This suggests that by this time weavers had already progressed from a circular warp to a flat warp that is wound onto the warp beam and gradually unwound as weaving progresses. A flat warp allows a much longer length of cloth to be woven. Flat wooden blades with pointed ends that

A World of Looms: Weaving Technology and Textile Arts

were found together with the warp beams might have been used as warp openers and weft-beaters, as in the Liangzhu loom. Other small pointed tools might have been used for selecting individual warps, for example for inserting patterning wefts.

We could reconstruct this loom as a type of simple body-tensioned ground loom.[3] Alternatively, we can note the similarity of the paddle shaped warp beams to those used on body-tensioned looms with a heddle attached to a long rocker, such as Tai and Miao frame looms in Southwest China[4] (cat. 15–17), as well as the traditional backstrap *bettle* loom from the Korean Peninsula (cat. 24) and the *jibata* domestic loom from Japan (sections 7.2.6 and 7.2.7). In this case the Jing'an looms may represent the earliest-known examples of cloth production on frame looms. It is common when weaving on this type of loom to remove the warp beam and cloth beam at the end of the day and roll them up together with the part-woven cloth. This removable set would probably have

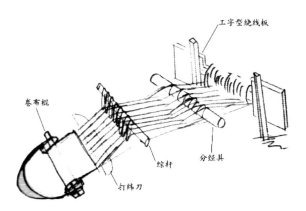

Reconstruction drawing of the Jing'an loom as a simple ground-level backstrap loom.

been considered sufficient to provide each person buried in the tomb with weaving equipment in the afterlife, and certainly would have been more convenient than placing an entire bulky loom in the coffin. (LB)

Loom used by Miao weavers in the Bakai district of Rongjiang county, Guizhou province. The paddle-shaped warp beam on the right is similar to that of the Jing'an loom (photo: Eric Boudot).

5 Shizhaishan Bronze Loom

Four loom parts made of bronze were excavated from Shizhaishan in Yunnan province. These parts were most likely used as a warp beam, a shed rod, a weft beater, and a cloth beam. Together they form a simple backstrap loom similar to that found in the Liangzhu tomb (cat. 3). The "shed rod" has a distinctive extra strip, forming an opening through which the warp passes, though such a rod could also have been used to make a heddle. This feature can also be seen on some looms used by Austronesian-speaking weavers in Taiwan, China, who are descended from weavers who migrated from the East Asian coast around 5,000–6,000 years ago. By this period a heddle would certainly have been in use because there were already more advanced frame looms—undoubtedly equipped with heddles—in nearby provinces.

The Shizhaishan site, as well as the Lijiashan site in Jiangchuan, also yielded bronze cowrie containers that depict textile workshop scenes (section 4, fig. 4.1). Both sites are dated to the Western Han Period (206 BCE–9 CE). The loom parts belong to the Dian Culture, a distinctive bronze-using culture with links to nearby Dong-Son Cultures in Guangxi Zhuang autonomous region in China and northern Vietnam that are known for the production of spectacular bronze drums. The textile workshop scene on the cowrie container shows several female weavers using foot-braced looms and a female overseer. This weaving scene suggests that the simple foot-braced loom had once been used widely on the East Asian mainland. (LB)

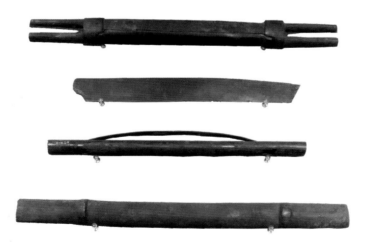

Bronze loom parts of the Dian Culture, excavated from Shizhaishan, Yunnan province. Collection of the Yunnan Provincial Museum. From the top: cloth beam, weft beater, shed rod or heddle rod, warp beam.

A World of Looms: Weaving Technology and Textile Arts

6 Luobowan Wooden Loom

A large number of weaving artifacts were excavated from tombs dated to the Han Dynasty at Luobowan, Guixian in what is now the Guangxi Zhuang autonomous region. Although the tombs were previously looted, the excavation still yielded many artifacts, including lacquer and bronze objects. Some objects are clearly textile weaving tools. When re-assembled, they form parts of several backstrap looms. Overall these loom parts are similar to those found in the Jing'an tomb (cat. 4) and probably correspond to the same loom type. (LB)

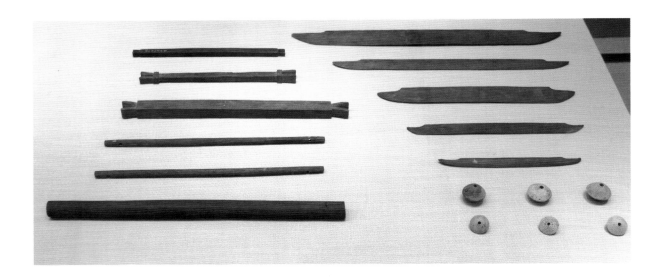

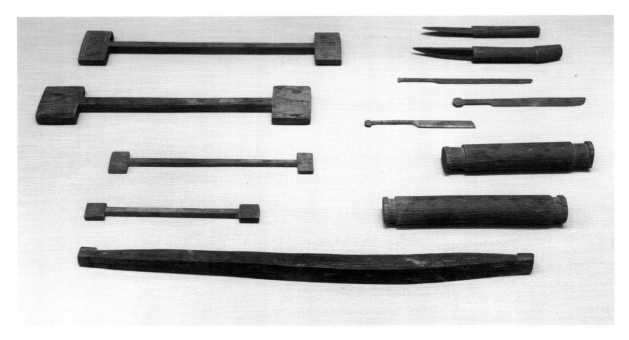

Wooden loom parts from the Han Dynasty, excavated from Luobowan, Guangxi Zhuang autonomous region. Collection of the Guangxi Provincial Museum.

7 Laoguanshan Pattern Looms

In 2012 and 2013, the Chengdu Institute of Cultural Relics and Archaeology and the Jingzhou Cultural Relics Conservation Center made a remarkable discovery at Laoguanshan in Tianhui, Sichuan province. While excavating four Han Dynasty tomb mounds containing wooden sarcophagi, they found a total of 620 objects made of lacquered wood, pottery, bronze, and iron. At the bottom of one wooden coffin they discovered four loom models made of wood and bamboo, along with traces of silk threads and pigments. The largest of the four looms is 50 cm tall, 70 cm long, and 20 cm wide. The other three looms are smaller, measuring around 45 cm × 60 cm × 15 cm. Of equal interest were the 15 miniature painted wooden human figures that, judging from their postures and inscriptions, may be models of the workers at a workshop where Sichuan *jin* silk was produced. To date these are the only complete models of Han Dynasty looms with firm provenance.

These looms are quite different than the looms that have been discussed so far. They are large and complex, and must have belonged to organized workshops producing high-quality silk textiles. In technical terms, the Laoguanshan model looms are "hook-shaft pattern looms," or "single hook and multi shaft pattern looms," which use two ground heddles and multiple patterning heddles to create patterns. The closest modern loom corresponding to this type is the *Dingqiao* loom (cat. 13), though the pattern-heddle opening mechanisms on the Laoguanshan looms are different from those on the *Dingqiao* loom. The four looms fall into two types: the largest loom is of the "sliding frame" type, and the other three are of the "linked shaft" type. The ground heddles are operated by a pair of foot treadles connected to a pair of pulleys in the upper part of the loom, and the pattern heddles are held in a cage-like frame, the position of which is fixed by a notched beam at the top of the loom. The

The Laoguanshan Han Dynasty tomb, Chengdu, Sichuan province, with loom models and attendant figures *in-situ*, visible on the lower part of the picture.

heddles are selected by a pair of suspended wooden hooks, and then raised by either the sliding frame or the notched beam via the foot treadles. The patterning heddles in the miniature looms number from 10 to 20. However, judging from actual silk textiles found in Warring States and Han Dynasty sites (5th century BCE–3rd century CE), actual full-size looms had many more heddles than the models.

Reconstructing the Laoguanshan Looms

The two types of pattern looms from the Laoguanshan tomb were reproduced as full-size looms in a collaborative project "Reproduction and Display of the Patterning Technology of a Han Dynasty Loom" carried out by the China National Silk Museum, in collaboration with the Chengdu Museum, the Institute for the History of Natural Sciences, Chinese Academy of Sciences, and the Zhijiang College of Zhejiang University of Technology.[5]

Following the reconstruction of the Laoguanshan loom, the NSM successfully reproduced a famous Han Dynasty *jin* silk using the hook-shaft pattern loom with sliding frames. The *jin* silk, which has been designated as a national-level cultural treasure, is an arm protector, probably used for archery (see p. 200). It was discovered from tomb 8 at the Niya site, Xinjiang. Bows and arrows have also been found in other tombs in the same area. The pattern is organized in a network of cloud scrolls that extend horizontally in the weft direction. Progressing from right to left are two birds, an animal with a single horn, and a tiger, interspersed with the characters *wuxing chu dongfang li Zhongguo* (five stars rising in the east are auspicious for the Middle Kingdom). Next to the inscription, there are two circular dots representing two of the Five Planets. The same tomb yielded another fragment of the same textile, depicting clouds, winged figures, planets, and three characters *zhu nan qiang*. (LB)

Model looms in-situ in the Laoguanshan tomb, with workshop helpers visible in the foreground (photo: Xie Tao).

East Asia: Archaeological Evidence from China

Reconstruction of one of the Laoguanshan looms at the China National Silk Museum.

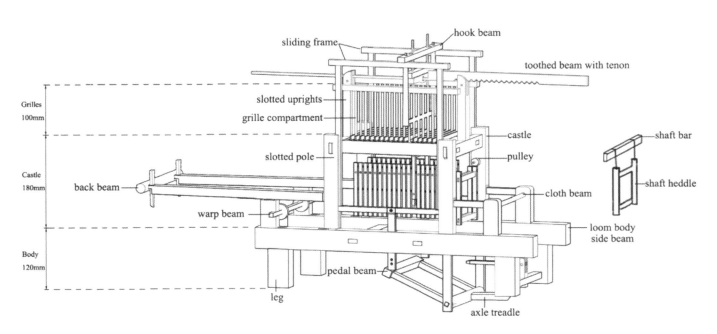

Reconstruction of one of the Laoguanshan looms, showing overall design and components (drawing: Long Bo).

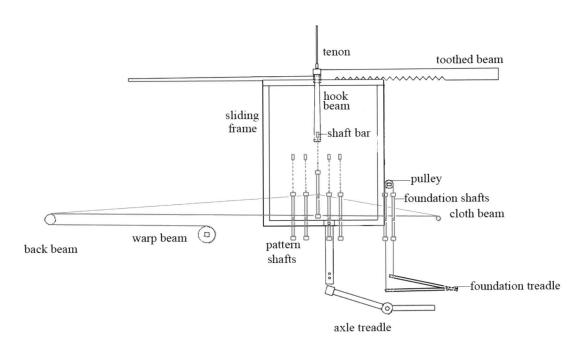

Reconstruction of the heddle lifting mechanisms (shafts) of one of the Laoguanshan looms. Red markings indicate the moving warp and shaft components of the loom (drawing: Zhao Feng and Long Bo).

Arm protector with *jin* silk excavated at the Niya site, Han Dynasty. Warp-faced compound plain weave; blue ground with pattern in yellow, white, green, and red. Pattern repeat in the warp direction: 7.4 cm; thread count: 220 warp x 24 weft ends per cm. Collection of the Xinjiang Institute of Cultural Relics and Archaeology.

Pattern reconstruction of the *wuxing chu dongfang li Zhongguo* silk, showing the position of the arm protector and an additional fragment from the same site (shaded areas).

Reweaving the *wuxing chu dongfang li Zhongguo* silk on the Laoguanshan loom.

8 Han Dynasty Oblique Treadle Loom

This loom was invented in China and widely used during the Han Dynasty. A key piece of evidence for this loom is a model made of green-glazed terracotta in the collection of the Musée Guimet in Paris. The form of the loom corresponds to the oblique treadle looms depicted on numerous stone relief carvings from the Han Dynasty. The pottery loom model shows a complete frame with a central axle, additional warp beam and pegs that indicate the positions of two treadles. There are two pairs of bearings on the slanting frame that probably supported the cloth beam and a shed opening. Although the treadles have been lost, two lugs for fixing the pedals are still present. The most important part is a pair of central pillars with short bars attached to them, probably the support for a heddle-lifting device. This exact type of loom has disappeared today, though some looms used by minority peoples in Southwest China use similar mechanisms for making warp openings (section 6.4). A number of reconstructions of this loom are possible, of which two principal types are reviewed here.

Han Dynasty pottery loom model (H. 30 cm; L. 25 cm; W. 17 cm). Collection of the Musée Guimet, Paris.

The first reconstruction is mainly based on the Musée Guimet model and on the stone relief illustrations. Its frame consists of two parts, a horizontal frame (横 机 身), and a slanting frame (斜机身). The horizontal frame acts as a seat for the weaver and a support for the slanting frame. The slanting frame provides the warp support and is inclined at about 60 degrees to the horizontal frame, hence the name of the loom. The main working components are attached to the slanting frame and, from top to bottom, include the warp beam, central axle, shed bar, "horse-head" board (马头) which supports a lever, and finally the cloth beam. The most important part of the loom is the shedding mechanism, which consists of the central axle, shed bar, "horse-head," and two pedals. Besides this it is assumed that a reed and temple were also employed in the loom.[6]

The working of the shedding mechanism in this reconstruction is as follows: the key part of this mechanism is the central axle, which has one pair of wooden "hands" (手子) attached to it at right angles. The "hands" are connected to the pedals separately. When the weaver depresses one pedal, the pedal pulls down one hanging "hand" (垂手子), and the axle turns in one direction, e.g. clockwise when viewing the loom from left. When the weaver depresses the other pedal thus pulling the "cross hand" (引手子), the axle will turn in the opposite direction. On the same level as the "hanging hand," there is a pair of "holding hands" (掌手子), which are linked to the ends of the levers, the other ends of levers linked to the shaft via a pair of wooden "elbows." One shed is opened by the shed bar naturally as the pedal pulls the "cross hand" and the central axle turns in the counter-clockwise direction. But when the other pedal pulls the "hanging hand" and turns the central axle clockwise, the "holding hands" will pull the wooden elbows that then pull the ends of the levers. And the other ends of the levers will lift the shaft to form a shed. In this way the warp on the slanting frame can be interlaced with the weft.

Some tomb murals seem to show two pairs of wooden "hands" attached to the central axle of the loom at a right angle, which will cause the axle to

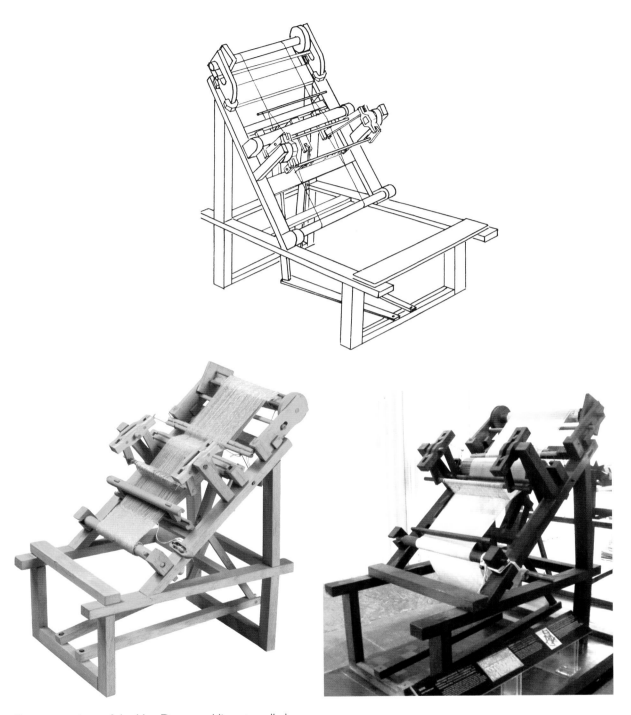

Reconstructions of the Han Dynasty oblique treadle loom.

rotate in the opposite direction. Obviously these looms have one central axle and two pedals. But others have pairs of wooden "hands" that turn in the same direction. These maybe belong to another type of loom with two central axles and two pedals, in which the mechanism has the same function as the loom described above.

An alternative reconstruction of this loom has been published by Boudot and Buckley.[7] This reconstruction is based on observation of a present-day loom that is used by Miao weavers in the Zhaoxing area of Liping county in Guizhou province, as well as some domestic looms with a fixed cloth beam that had been used in China until

205

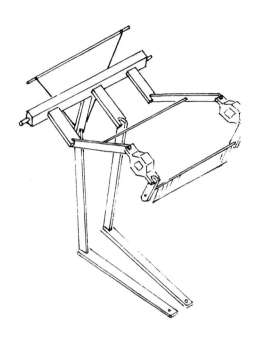

The axle-treadle mechanism of the Han Dynasty oblique treadle loom.

the mid-20th century. The frames of these looms are somewhat different from the Han Dynasty loom, but the shed opening arrangements offer clues to how the Han loom might have operated. There are two treadles: one is attached to a pair of short rockers (corresponding to the central pillars with short bars in the Musée Guimet loom model) and then to the heddle. The other treadle is attached to a shed stick that sits in the natural shed. It is used for restoring the natural shed opening. A single treadle is sufficient for opening and closing the counter-shed on looms that have body tensioning, because the weaver is able to lean back and increase the warp tension to restore the natural shed, but in looms with a fixed cloth beam she cannot do this, and the heddle tends to "stick" at the half-way position. The extra treadle linked to the shed stick helps to overcome this problem, pulling the warp back down and reopening the natural shed. The treadles are then used alternately to open shed and counter-shed. (ZF and LB)

Alternate reconstruction of the oblique treadle loom (after Boudot and Buckley, 2015). The shaft (opening the counter-shed) is shown in green, the shed stick and its cords (re-opening the natural shed) are shown in brown.

9 Han Dynasty Vertical Treadle Loom

This loom is an adaptation of the oblique treadle loom and takes its name because the warp is perpendicular to the ground. A depiction of the loom appears in the murals from the Dunhuang grottoes, dated to the Five Dynasties Period (907–960).

During the period from the Han to Tang Dynasties, an evolution seems to have occurred from a slanting frame to a vertical frame, though the shedding mechanism remained basically the same. Both pictorial and textual depictions found at the Mogao Grottoes in Dunhuang show a vertical treadle loom. Several documents, the earliest dated 884 CE, mention *liji* (立 机) or *liju* (立 居), a kind of cotton fabric woven on a vertical loom. There is also a drawing in the wall paintings in cave 98, showing a vertical loom with a vertical frame and two pedals. A little later, in the Northern Song Dynasty (960–1127), a wall painting in the Kaihua Temple, Shanxi province depicts a vertical treadle loom operated by a female weaver.

The most detailed textual and pictorial depiction of a vertical treadle loom is found in the book *Ziren Yizhi* (《梓人遗制》) or *Traditions of the Joiner's Craft* by Xue Jingshi (薛景石) who lived in Shanxi during the Yuan Dynasty (1206–1368). The text records all the components of the loom, giving their shapes and dimensions, as well as the method of construction. Based on this description,

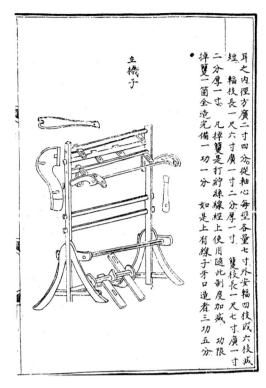

Depiction of a vertical treadle loom in the Yuan Dynasty *Traditions of the Joiner's Craft*.

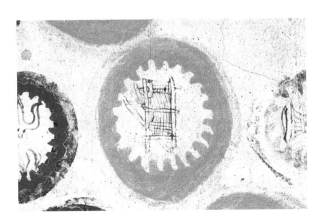

Depiction of a vertical treadle loom in the Mogao Grottoes in Dunhuang, Five Dynasties Period.

Illustration of a vertical treadle loom from a Song Dynasty mural.

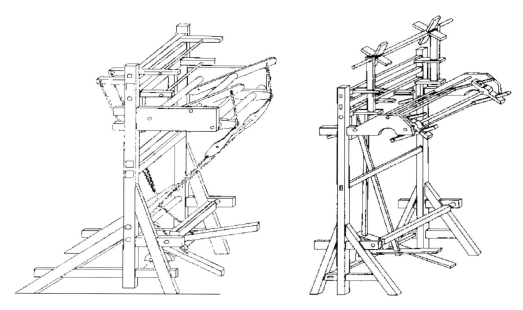

Reconstructions of the Yuan Dynasty vertical treadle loom by Dieter Kuhn (left) and Zhao Feng (right).

Zhao Feng has made a replica of this loom and woven a piece of plain weave fabric using it. This loom is also a kind of axle-treadle loom with a number of ingenious mechanisms. Two pedals can pull and push the "cross hands" to drive the central axle in different directions through the rigid arms. The rotating central axle controls the levers and makes the single shaft lift or fall back to open shed and counter shed. A notable point is that the warp beam of this vertical-axle treadle loom can be moved to maintain the warp tension: when the group of warp ends, pulled through the heddle, is lifted and the warp tension is increased, the warp beam, located on the top of the loom, is lowered and the tension is decreased, and when the warp loosened then the warp beam pushed upwards and the tension increased.[8]

These features, with the exception of movable warp beam, are very similar to those of the oblique treadle loom in the Han times. Such looms also appear in the Ming Dynasty (1368–1644) paintings such as *Imperial Palace* (《宮蠶圖》), which depict silk weaving. Several years ago a true vertical treadle loom, apparently the same as that recorded by Xue Jinshi, was seen in a small village in southern Shanxi province.

It seems clear that the vertical treadle loom

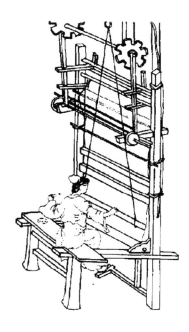

Depiction of silk weaving in the Ming Dynasty *Imperial Palace*.

evolved out of the oblique treadle loom, but it does not seem to have ever been as widely used as the latter. Most of the depictions, from the Song Dynasty to the present, were found in or near to Shanxi province, so we might conclude that Shanxi was the last place where the axle treadle loom was

East Asia: Archaeological Evidence from China

Working reconstruction of the vertical treadle loom by Zhao Feng.

in regular use. It also seems that the axle-treadle loom, although it continued to develop after the Han Dynasty, gradually declined importance in the succeeding dynasties. In its role as a non-patterning loom, it was replaced by the lever treadle loom, first by one with two shafts and two treadles, the independent lever treadle loom, so called because of the independent motion of the two levers and the shafts, and then by the balanced (linked) treadle loom (double heddle loom), the most popular type of loom from the 13th century until the present time. (ZF and LB)

209

East Asia: Present-Day Looms from China

10 Backstrap Treadle Loom

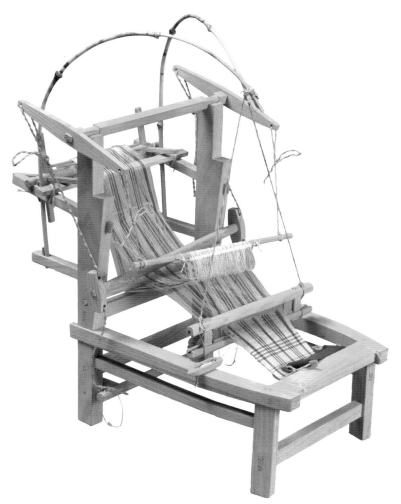

Backstrap treadle loom, with partly-woven cloth.

The use of a single treadle to operate the heddle for alternating sheds is a distinctive feature of a backstrap treadle loom. This loom is an advance on the simple backstrap loom. While the cloth beam is still connected to the waist of the weaver, the warp beam is now fixed to a frame. This type of loom first appeared during the Han Dynasty and was in widespread use in rural villages in China until the latter part of the 20th century, for making rolls of plain cloth, as well a stripes and checks. It continues to be used by some minority groups in Southwest China (section 6.4).

The main feature of this loom is that a single heddle is attached to a short rocker, attached to a treadle via cords. When the short rocker turns, it lifts the heddle rod and opens the counter-shed.

Ramie cloth, of a type that was formerly woven in central and southern China on backstrap treadle looms and then dyed with indigo using the paste-resist technique, mid-20th century. Private collection.

The natural shed is retained by an opening in the warp created by the loom frame. The "shaft" (heddle-plus-treadle) in this loom was a major technological advance.

During the Han Dynasty, although most of the non-patterning looms depicted in tomb murals can be considered axle-treadle looms, there is one mural from a Han Dynasty tomb at Zengjiabao, Sichuan that seems to be a loom of this type.[9] Some looms of this type, including the one illustrated here, have a pair of curved bamboo rods that suspend the reed in the warp for ease of use. (LB)

11 Balanced Paired-Treadle Loom

Balanced paired-treadle loom.

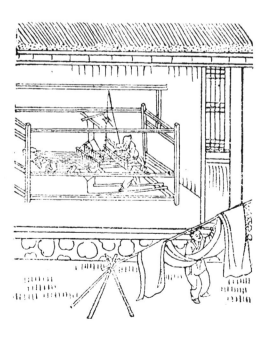

Balanced treadle loom depicted in the Qing Dynasty *Collection of Important Essays on Sericulture*.

Most of the looms mentioned so far are unique to China or to East and Southeast Asia. The "Balanced Paired Treadle Loom" however is the China representative of the double-heddle loom that is now used worldwide, with many different forms and variants. This loom employs foot treadles that raise and lower a pair of clasped (bidirectional) heddles to open two wide sheds in a symmetrical fashion. This system marks a major technological advancement in loom evolution, and this loom has been gradually replacing older types, becoming widespread in the Qing Dynasty (1616–1911). Both pictorial and textual depictions can be found in many historical documents, such as the *Can Sang Cui Bian* (*Collection of Important Essays on Sericulture*) of the Qing Dynasty, showing a weaver working on a balanced treadle loom. Versions of this loom can be found in villages in most parts of China, and it is an efficient way to produce plain weave cloth. (LB)

12 *Kesi* Loom

Kesi loom.

Kesi or slit-tapestry is a weft-faced textile. It is woven with un-dyed warps and multi-colored discontinuous wefts that serve both as the ground-weave and as the pattern. Weaving with each color weft is done following the shape of a particular motif as opposed to a strict order of the shed, row by row. Tapestry was first woven with wool in the west of China, the technique being later adapted to silk in China during the Tang Dynasty (618–907) and gaining popularity in the Song Dynasty (960–1279). It was used for producing complex designs for imperial robes and religious images. *Kesi* weaving on looms such as the one shown here still continues to this day. (LB)

The border of a Qing Dynasty imperial robe, patterned with waves and auspicious symbols, woven in *kesi* technique. Private collection.

13 *Dingqiao* Loom

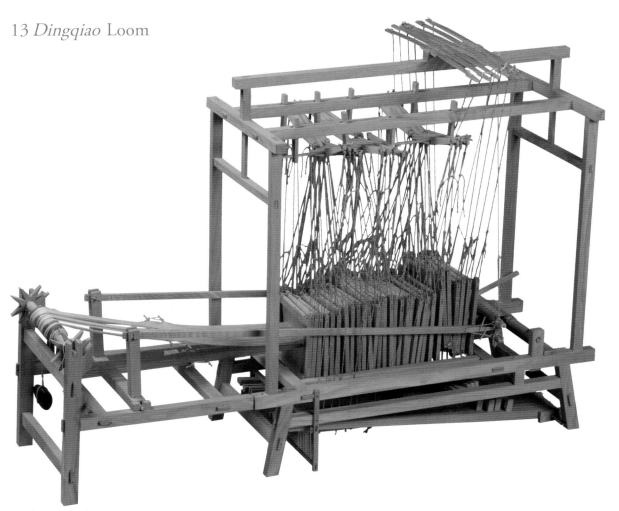

Dingquiao loom.

The "stepping stone" treadles that give the loom its name.

The *dingqiao* or "stepping stone" loom is a multi-shaft and multi-treadle pattern loom. It gets its name from the small blocks on the treadles that look like stepping stones crossing a river. It is used to weave ribbons in the villages near Chengdu, Sichuan province, and it was formerly used to weave wider lengths of silk with short pattern repeats. The loom usually has 2–8 depression shafts for the ground weave and 40–60 lifting shafts for the pattern. Each shaft is controlled by a treadle; therefore, the number of shafts equals the number of treadles. This type of pattern loom has a long history that goes back to the 3rd century, attested by historical documents from that period. (LB)

14 Taiwan Atayal Loom

Versions of this backstrap loom are used by minority groups in Taiwan speaking Austronesian languages. The loom is a rare survival of an ancient type of loom, similar to those found in archaeological remains of the Liangzhu Culture (cat. 3) and the Dian Culture in Yunnan (cat. 5), as well as weavers in a few isolated regions such as Hainan and the Lao-Vietnam border regions (sections 6.1 and 9.1). The cloth beam is secured at the weaver's waist, while the large, hollow warp beam (on the right in the photograph) rests behind the weaver's feet. This large, hollow warp beam is a unique feature of Taiwan looms. To adjust the tension in the warp the weaver flexes the warp beam with her feet. Fibers for weaving include hemp, ramie, cotton and hibiscus. Striped patterns and small geometric designs are made using various aids, including heddles and sticks embedded in the warp. Weaving on this type of loom had become rare, but recently there has been a revival of interest in traditional weaving in Taiwan, with many young weavers now learning the craft. (CB)

Above: a weaver from Taiwan demonstrating the Atayal loom. The large warp beam (right) is braced by the weaver's feet. Below: the complete loom.

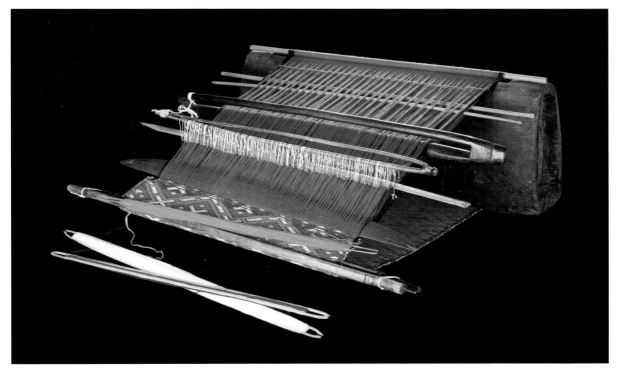

15 Dong (Kam) Brocade Loom

Known as a *douji*, this loom is unique to the Dong people living in the Tongdao area of Hunan province and parts of nearby Guizhou province. They use it to weave cloth decorated with supplementary weft, for use on baby carriers and wedding bedcovers. As with all body-tensioned looms in this region, the warp beam is fixed in the loom and the cloth beam is attached to the weaver's waist, so this is an example of a "half-frame loom." The weaver has two loops of cord around her feet: pulling one raises the ground-weave heddle, while pulling the other raises the pattern heddle. The pattern heddle hangs down in a "curtain," but it is actually circular and is used in a similar way to the Zhuang bamboo cage loom (cat. 17), to which it is closely related. Pattern sticks are used in sequence and re-inserted after use, as with other patterning looms of this general type. The ground weave is usually cotton, while the colorful wefts can be either cotton or silk.

As with many other weaving groups in Southwest China, the Dong people also use looms of the double-heddle type to weave plain cotton cloth, but they keep this older type of loom for making cloth with supplementary weft patterns. This conservative habit has ensured the survival of some very ancient types of loom in this region, despite the advances made by more modern looms. (CB)

Above: Dong brocade loom. Facing page: detail of a wedding bedcover woven on a Dong brocade loom. Cotton, white and indigo-dyed. Private collection.

16 Dai Jinghong Brocade Loom

This distinctive loom is used by Dai weavers in the Jinghong region of Yunnan province. As with the Dong loom and Zhuang loom from the Binyang region, the cloth beam is attached to the weaver's waist, and the warp beam is lodged in the loom frame. There are two treadles: depressing one raises the heddle that opens the ground weave shed, while depressing the other raises the pattern harness. The basic principle of operation is similar to the Zhuang bamboo cage loom, though the arrangement of leashes is different. Patterns are stored on looped threads or sticks in a pattern harness that stretches above and below the loom: each loop records one pattern weft insertion. After each pass of the weft, the "active" loop, is transferred from the top of the pattern harness to the underneath part. A loom may have as many as 100 looped threads. A similar patterning system is used by Tai weavers in Vietnam, Laos and Thailand, though the looms used by these weavers are different (cat. 26): they are not body-tensioned, and the patterning system is fixed to the loom frame rather than a treadle-and-rocker system as in the Tai loom. (CB)

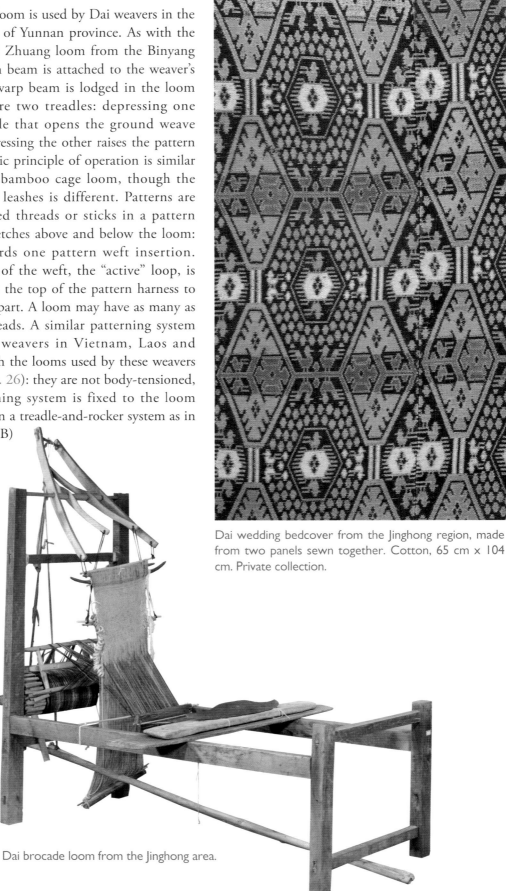

Dai wedding bedcover from the Jinghong region, made from two panels sewn together. Cotton, 65 cm x 104 cm. Private collection.

Dai brocade loom from the Jinghong area.

17 Zhuang Binyang Bamboo Cage Loom

Zhuang people in Binyang region in Guangxi use a backstrap treadle loom with the ground-weave heddle attached to a long rocker, a type of loom that is used by many minority peoples in Southwest China. The unique feature of Zhuang looms from this region, as well as other Tai looms of the Tai and Dong for example (cat. 15 and 16) is a complex pattern heddle, in this case built around a large bamboo cage, which stores the patterns for supplementary weft designs. Around 100 patterning rods made of bamboo are embedded in a single set of leashes, held around the cage. The whole cage assembly is attached to a second treadle, so the weaver can raise and lower it to bring it into position. She uses each of the patterning rods in turn, pulling each one forward to raise warps, then inserting silk patterning wefts. After using the rod she removes it and re-inserts it in the pattern leashes behind the drum, preserving the design for repeated use. In essence this is the simplest kind of drawloom system, recording warp lifts in a circular sequence that can be re-used endlessly. The main difference between this loom and (for example) the Greater Drawloom used in Han Chinese silk workshops (cat. 22 and 23) is that the latter has an extra set of leashes between the patterning system and the warps, and the option to connect each pattern leash to multiple warps to make pattern repeats, whereas in Tai looms one pattern leash is always connected to one warp. The Greater Drawloom also incorporates many more pattern rows, which are encoded using cords rather than sticks. (CB)

Zhuang brocade loom from the Binyang area.

18 Zhuang Jingxi Brocade Loom

This loom is characteristic of Zhuang weavers in Jingxi county in northwestern Guangxi Zhuang autonomous region. They use it to make their distinctive bedcovers that are decorated with silk supplementary weft. The Zhuang are a Tai group, related to other Tai peoples in Southwest China and Southeast Asia, including the Dong (Kam), Maonan, Shui, Dai and Buyi in China, Tay in Vietnam and Thai/Tai in Vietnam, Laos and Thailand. Many Tai groups weave intricately decorated cloth with silk supplementary weft decoration for a variety of purposes, including wedding bed covers, skirts, shoulder cloths and headscarves. They use a variety of different frame looms for this purpose, many of which include complex patterning systems.

Detail of weaving by the Zhuang people from the Binyang area. Silk supplementary-weft patterning on cotton ground. Private collection.

Zhuang loom from the Jingxi area, with supplementary weft heddles which record the pattern supported by red cords in the center of the loom.

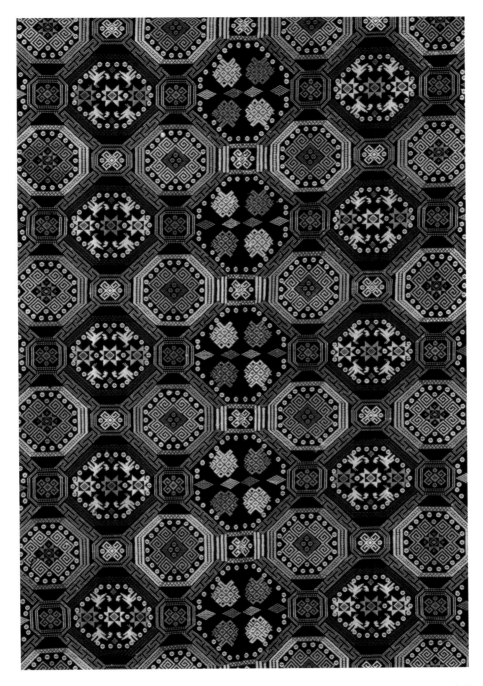

Wedding bedcover made by the Zhuang people from the Jingxi region. Silk supplementary-weft patterning on cotton ground.

In this loom both the warp beam and cloth beam are fixed in the frame. The ground-weave sheds are opened with a pair of treadles and heddles (in other words, this is balanced treadle or double-heddle loom). The warp lifts for pattern weft insertions are stored on individual heddles on a rack above the warp, arranged in sequence. The weaver picks out each heddle and lifts it by hand, inserting a bamboo tube into the warp to hold the shed open before inserting a colored silk weft. A heavy reed attached to a swinging arm is used to beat in the wefts firmly. When she has completed a set of weft insertions she changes direction and uses the pattern heddles in reverse order, creating a symmetrical design. (CB)

19 Velvet Loom

The basic shape of this loom is related to a drawloom, though strictly speaking it is not a drawloom since it only has ground-weave heddles and lacks a patterning tower. The weaver manipulates two warps: the foundation warp and the velvet warp. The latter is pulled up in short loops to form the design, thus it has a much longer take-up than that of the foundation warp. Because these warps have different take-up, each requires a separate warp beam.

Velvet is woven in plain weave with loops of warp sticking up to make the pile, which are formed around thin metal rods. Several different types of loom were formely used for its manufacture, depending on whether the velvet was patterned or not.

The pile loops can be cut to produce a soft fuzzy surface, or left uncut (section 15.3, fig. 15.4). The weaving process requires a high degree of skill. A textile excavated at tomb I at Mawangdui in Hunan province shows that loop-pile polychrome *jin* fabric was already being produced by the 2nd century BCE. In more recent times Zhangzhou in Fujian province has been regarded as a center for velvet production. (LB and CB)

Above: velvet loom. Facing page: silk velvet with gold-thread supplementary-weft patterning, Qing Dynasty. Private collection.

20 Damask Drawloom

Damask is a monochrome textile in which the pattern is formed by contrasting weave textures, making a lustrous and shimmering surface effect. The loom for weaving damask belongs to a type of drawloom called the Horizontal or Lesser Drawloom.

Records show that damask fabrics formed part of the tributes to the imperial court as early as the Tang Dynasty. Such damasks would have been woven on this type of drawloom using the local *jili* silk. The late Tang Dynasty and Song Dynasty writings discovered at Dunhuang frequently mention ling damask made on a tower loom (*lou ji ling*), suggesting that drawloom was already being used to weave such type of silk during that time. This particular loom comes from the Shuanglin region near Huzhou city in Zhejiang province, which is famous for the production of high-quality damask. (LB)

Above: damask drawloom. Below: silk damask with dragon and clouds design. Qing Dynasty. Private collection.

21 Gauze Drawloom

This loom is a type of Lesser Drawloom that has been adapted for making gauze. It is equipped with two types of harnesses: standard harnesses and special doup harnesses that raise alternate warps, moving them to one side and back again, creating twists in the warps, through which the weft is inserted. The twists in the ground weave create a stable structure that can have large openings, which allow weavers to produce a light and cool fabric. The loom has a pattern tower controlled by an assistant (drawperson), who is responsible for adding damask-like repeating patterns to the finished textile. (LB and CB)

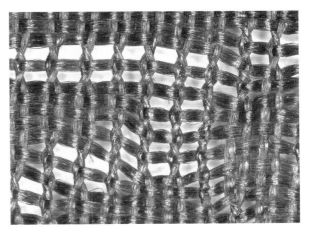

Macrophotograph of a patterned silk with gauze weave, showing crossed warps in the patterned areas. Qing Dynasty.

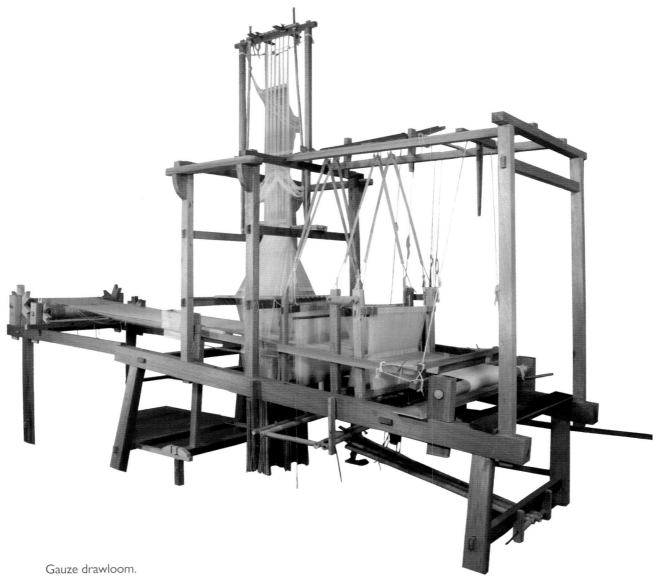

Gauze drawloom.

22 Lampas Drawloom

The drawloom for making lampas is a type of Greater Drawloom. It is similar to the drawlooms for *yun* brocade (cat. 23), but it has a set of harnesses that are specific for lampas weave structure. In lampas weave, the pattern wefts are bound by an extra set of warps, called the binding warps. The ground weave is normally warp-faced, while the pattern is weft-faced.

The term "Song lampas" refers to a luxurious type of silk textile that was popularly used for court costumes and for mounting paintings and calligraphy. It imitates Song Dynasty silk in patterns and colors, but in some cases its weave structure is actually a weft-faced compound weave instead of lampas. At the time of the Southern Song Dynasty, there were more than 40 different patterns. In

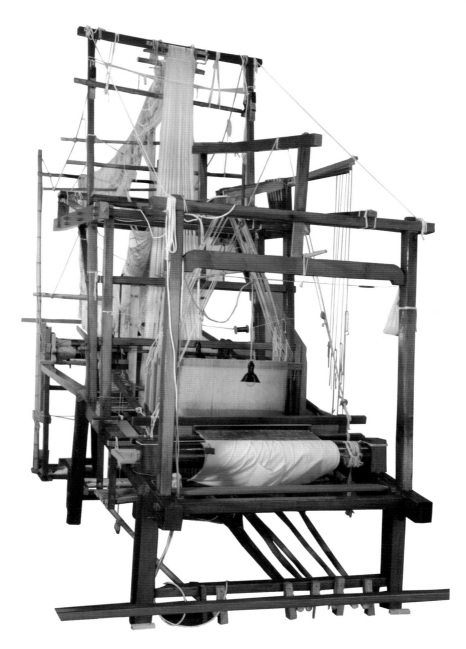

Above: lampas drawloom. Facing page: silk lampas from the Ming Dynasty. Colorful supplementary-wefts float over a warp-faced ground weave of green silk. About 2x actual size. Private Collection.

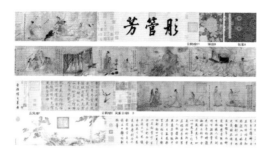

Above: the *Admonitions* scroll, attributed to Gu Kaizhi (c. 345–406), showing part of the lampas weave silk that decorates the cover. Collection of the British Museum, London.

Below: modern replica of the silk lampas cover woven at the NSM.

23 *Yun* Brocade Drawloom

This loom is a variant of the Greater Drawloom, with ground weave heddles and patterning system of similar design and complexity to the lampas drawloom (cat. 22). This loom was used to weave a large variety of polychrome silks that were used for making imperial robes and furnishings. Some of the most common designs on these polychrome silk "brocade" include dragons, phoenix, and clouds (*yun*), from which the *yun* brocade loom received its name. These silks were important not only in China; as precious items, they were also presented as diplomatic gifts and traded in neighboring regions. The designs of these silks influenced many textile traditions across East and Southeast Asia. (LB)

the Ming and Qing Dynasties, Song lampas was classified into three types, depending on the size of the patterns.

The blue silk lampas with a lattice pattern in the illustrations above is a replica of a textile on the cover of the famous painting *Admonitions of the Instructress to the Court Ladies*, an 8th-century copy of one of the earliest and finest paintings in China. The same textile is found on the covers of three other Song Dynasty scroll paintings: *Nine Songs*, *Shu Rivers*, and *Imaginary Tour Through Xiao-Xiang*. Silks with similar patterns—based upon variations on a lattice made up of squares and octagons—were produced in large number from the Song Dynasty until the late Qing Dynasty. These silks would have been used on scroll painting covers and book boxes. The textile replica was woven on the lampas drawloom in the China National Silk Museum. (LB)

Silk with design of dragon and clouds, Qing Dynasty. Private collection.

East Asia: Present-Day Looms from China

Yun brocade drawloom.

Silk with elephant design, Ming Dynasty. Private collection.

229

Additional Weaving Tools

In traditional looms, tools for inserting wefts and beating-in wefts are also important parts of the loom. The simplest weft-insertion tool is a short wooden or bamboo stick around which the weft is wound. A slightly more sophisticated version is the spool, a long rod with notches at each end that retain the yarn. Yet another type is a shuttle, the most common weft insertion tool on a frame loom, its boat-like shape seems to have found its present-day form during the Warring States Period. A version of the shuttle that sets the weft yarn inside a beater can be used for both weft insertion and subsequent beating-in, and is common on domestic looms.

There are three main methods of beating-in wefts, an essential step for making a firm weave. One is to beat the weft with the edge of a weaver's "sword," an elongated wooden blade of some kind, usually thin and shaped like a sword. The second method is to use a fork-like device, such as a comb. A simple comb-beater is used in *kesi* and carpet weaving. The third method is to use a reed, where the comb shape encloses the warp. Reeds have been widely used since the Qin and Han Dynasties. The reed has the advantage over other types of weft beaters in that it also helps to regulate the width of the fabric. (LB)

Shuttle.

East Asia: The Korean Peninsula

24 Korean *Bettle* Backstrap Treadle Loom

This type of backstrap loom, called *bettle*, has a long history on the Korean Peninsula, and is still used for making plain weave textiles in silk, hemp, cotton, and ramie. Occasionally, it is also used to make ribbed gauze fabrics. The *bettle* loom is constructed using three components: a main frame that slopes upward from the weaver, an I-shaped warp beam, and a cloth beam that is connected to the waistband of the weaver. The natural shed is held open by a small triangular frame, while the counter shed is formed by a heddle attached to a rod. For weaving ribbed gauze the loom would have two heddle rods rather than one.

The *bettle* has two key characteristics. The first is the way the heddle rod is lifted. When the weaver pulls the rope with his or her foot, this action causes a connecting wooden bar to rotate, then a long rocker to rise, which pulls up the heddle. The second is the triangular frame that divides the warp threads to form the natural shed. This component is not fixed to the frame; thus, a weaver can manually insert or withdraw it from the warp. The *bettle* is widely used on the Korean Peninsula. Related looms are used in southern parts of China (section 6.3, cat. 15–17) and parts of Japan, where it is known as the *jibata* loom (sections 7.2.6 and 7.2.7). (SY)

Bettle loom.

Mainland Southeast Asia

Introduction

The following catalog entries on mainland Southeast Asian looms focus on frame looms used by Tai and Khmer weavers, mainly for weaving silk textiles decorated with supplementary weft and ikat. These are used to make some of the most accomplished and luxurious weaving in the region. Some looms employ pattern saving systems to record designs, while in other cases weavers pick out designs by hand.

A Tay weaver in Ban Rang village in northern Vietnam weaving cloth for a bed cover, decorated with supplementary weft in bright colors, which hang down in butterfly loops below the working area. The warp lifts for the patterns are recorded in sticks around the small drum suspended in front of her (photo: Christopher Buckley).

25 Tai Frame Loom

This loom was collected in Phu Phan, Sakon Nakhon province, Thailand. Women from Tai groups such as Lao and Thai and related groups such as Tai Phuan and Phuthai weave on frame looms consisting of four posts and cross-beams to keep the frame upright. The warp beam, reed, and heddles are strung from rope to wood or bamboo poles placed on top of the frame. A rope connects a pair of treadles to the main heddles (1:1). The cloth beam is anchored to short posts in front of the seat. This frame loom is used to weave weft ikat plain weave. A weaver may install a supplementary heddle system to apply continuous or discontinuous supplementary-weft technique, and additional principal heddles can be incorporated to create twill weave. (LM)

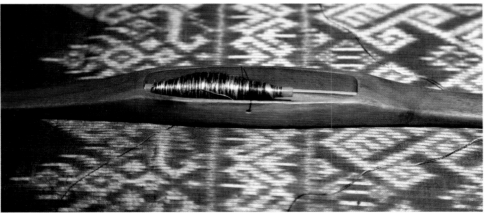

Above: loom from Phu Phan, Thailand. Below: detail of weft ikat on the loom, and a wooden shuttle.

A tubeskirt woven by a Tai Phuan weaver, made from silk decorated with weft ikat, with a blue cotton waistband. Private collection.

26 Lao Tai Frame Loom with Vertical Pattern Heddle

Gift of Mary Connors.

The Lao Tai groups use variations of a frame loom to produce complex textiles of tapestry, twill and plain weaves. Their cloth may be embellished with supplementary warp threads, as well as continuous and discontinuous supplementary weft threads. They have developed a system with separate ground weave and pattern weave heddles to create and preserve complex supplementary designs, incorporating a pattern "code" that can be used and reused. Warps are threaded through the secondary long-eyed heddle system and can be raised or lowered depending on the pattern. The pattern is stored in the vertical heddle, with each row of weft insertion recorded by a bamboo rod or string. This technology allows the weaver to store and preserve her creative expression realized in her laboriously picked design. As an important part of her heritage, it may stay with her when she marries or travels. (CC and IL)

Lao Tai frame loom.

A World of Loms: Weaving Technology and Textile Arts

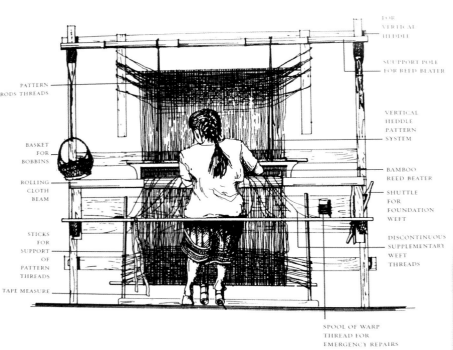

Above: Schematic drawing of the Lao Tai frame loom with vertical heddles (Carol Cassidy).

Right: Coffin cover or shroud. Silk, decorated with continuous and discontinuous supplementary weft. Tai Daeng group, Hua Phan province, northeast Laos, early 20th century. Gift of Mary Connors.

Mainland Southeast Asia

Woman's ceremonial skirt (central section), woven from silk. Tai Daeng people, north Lao-Vietnam border region. Skirts of this type, which were worn mainly for weddings and funerals, are a *tour-de-force* of decorative techniques applied on the loom. The wide bands are decorated with weft ikat designs of stylized dragons (*naga*) and other motifs. In between there are blocks of brightly colored silk supplementary weft, and bands of warp patterning. The warp runs horizontally in this photograph. Gift of Mary Connors.

27 Tai Phuan Frame Loom

These are the heddle system, reed and treadles of a loom from the Tai Phuan group, Sri Satchanalai, Sukhothai province, Thailand, collected in 1989. They would have been mounted on a frame loom of similar type to cat. 25.

A weaver would use a reed of a narrow width, such as this one, to create skirt borders. The borders of ceremonial skirts were often decorated with discontinuous supplementary technique (*chok*). She would "pick out" the pattern by hand or use pattern sticks placed in the warp or in a vertical, supplementary heddle system (cat. 26) to save the design. Tai Yuan weavers of Thailand apply this technique with the "right" side of the pattern facing upwards. Tai Phuan weavers of Sri Satchanalai, Sukhothai province whose ancestors moved to this area from Laos in the early 19th century subsequently adopted the Tai Yuan method of creating *chok*. (LM)

Woman's tubular skirt, *sin muk tiin chok:* silk, cotton, supplementary warp *muk* (mid-section), discontinuous supplementary weft *chok* (border). Tai Yuan group, Lap Lae, Uttaradit province, Thailand, early 20th century. Private collection.

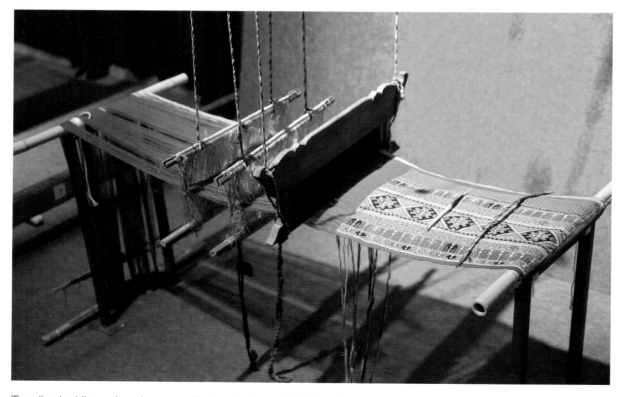

Treadles, heddles and partly woven skirt border from a Tai Phuan loom.

28 Khmer Frame Loom

The weft resist-dyed or ikat decorated silk textiles woven by Khmer and Cham weavers in Cambodia are renowned for their fineness and intricacy. These weft ikat silks (*sampot hol*) are distinct from other weft ikats in the region since they are produced in a 2:1 twill weave. The loom illustrated here was developed for high-quality commercial production; additional equipment assists in keeping threads aligned. Four heddles located behind the comb or reed indicate that the loom is set up for weaving twill such as the diamond-shaped *lboeuk* pattern. The 2:1 twill weave that traditional weft ikat silks requires three heddles. Producers are combining weft ikat technique with a 4-heddle twill weave, such as with the *lboeuk* design, for commercial production. (LM)

Detail of ceremonial textile, *pidan hol*: silk, weft ikat, 2:1 twill weave. Khmer people, Cambodia, mid-20th century. *Pidan* means "ceiling" in Khmer, and textiles illustrating episodes of the historical Buddha's lives are hung as canopies above Buddha images in temples, or as wall hangings inside or outside temples during rites.

Khmer frame loom, with *pidan hol* textile behind.

Insular Southeast Asia

Introduction

Weaving in Island Southeast Asia is mainly done on body-tensioned looms located at ground level. There are many different variations, of which a few are shown here. The simplicity of the looms does not limit the complexity of weavings, which include luxurious textiles of great intricacy and beauty.

29 Sulawesi Body-Tensioned Loom for *Tenun Mamasa*

This loom was owned by Limbongtata, a weaver from West Sulawesi. The partly woven textile on the loom is made from polyester yarns dyed with synthetic dyes, around 10 m long and 40 cm wide. It is a traditional hip wrapper used in ceremonies in Mamasa, mountainous area in West Sulawesi; thus, it is commonly called *tenun mamasa*,

Weavers in Mamasa, weaving cloth on externally braced, body-tensioned looms. Cloth beams are fixed to the weaver's waist, warp beams are attached to house supports. Similar looms are used across Insular SEA (photo: Dinny Jusuf).

meaning "weaving from Mamasa." This type of textile is made in plain weave, often featuring contrasting stripe patterns that are produced by a simple arrangement of several colored warps, as well as small motifs created by a warp-float technique called *sakka*. The *sakka* motifs are often derived from cultural objects such as the traditional houses (*tongkonan*) and blades (*gayang*). They also take inspiration from the local scenery and animals such as mountains, crabs, etc.

Weaving is usually done outdoors. A weaver would sit with legs fully extended on a raised platform under a rice barn. One end of the warp yarns would be tied to a post, which acts as a warp beam. The other end would extend to a cloth beam on the weaver's waist, and the cloth beam would connect to a wooden yoke behind the weaver's back through a strap. The weaver would control the tension of the warp by leaning forward and backward. In the past, *tenun mamasa* could be made from many types of plant fibers such as cotton, pineapple, and other various plants that grew in the surrounding area. However, since around the late 1960s—with the growing influx of synthetic materials and the diminishing natural resources—Mamasa weavers stopped using natural materials. (DJ)

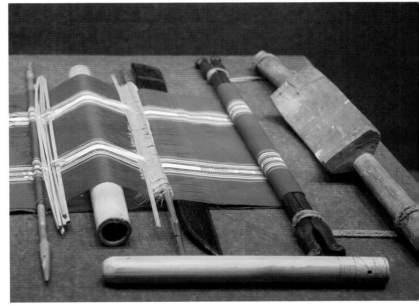

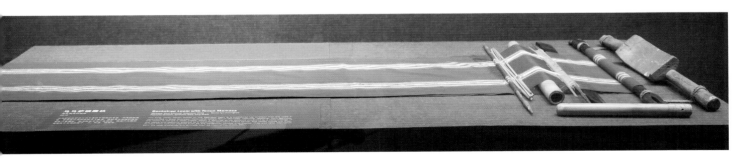

Above left: *tenun mamasa* worn as hip wrapper (photo: Dinny Jusuf).
Above right and below: detail and overall of loom for *tenun mamasa*. Collected by NSM in 2018.

30 Bali Body-Tensioned Loom for *Geringsing* Double Ikat

The NSM commissioned this loom in 2018 from Ni Luh Kembang, an experienced weaver in Tenganan Pegringsingan village, Bali. Tenganan Pegringsingan is the only place in Indonesia where double *ikats* are made. Called *geringsing*, these textiles serve as ritual clothing in various ceremonies. *Geringsing* alludes to the name of the village where the textiles come from, but also to the words *gring*, meaning "sickness," and *sing*, meaning "no," i.e., "no sickness."

Geringsing textiles are made of loosely woven cotton threads and natural dyes. There are three colors: yellow from candle-nuts, brownish-red from the roots of *Morinda citrifolia*, and blue from indigo leaves. The dark blue-black color is produced by the overdyeing of red and blue. Each color signifies an element: yellow for wind, red for fire, and black for water. The villagers believe that these three elements are present within each person, and when they coexist in equal balance, the body will be healthy.

The double ikat technique requires both the warp and the weft to be tie-dyed prior to weaving. To achieve a clear design, the weaver has to carefully align the tie-dyed patterns on these two sets of threads as the weaving progresses. There are several traditional patterns, which can be combined in various ways. The half-woven textile on the loom displays two patterns, *cemplong* and *lubheng*. A full-width textile with elaborate designs displays the patterns *wayang putri* on the main body and *cempaka* at its ends.

The *wayang* pattern is considered to be the most complicated *geringsing* design. *Wayang*, meaning shadow puppets, refers to the two figures—shown as double in mirror image—within each half-circle layout. Four of these half-circles connect to form a cross-like imagery; to the villagers, this crossing is a reminder to maintain the balance of their lives.

The *geringsing* loom and textiles on display were produced through the collaboration of several individuals within an extended family. The weaver's husband provided the aged wood for the material of the loom, his friend made it, and the weaver and her mother both dyed and prepared the ikat for the half-woven textile. An aunt of the weaver wove the *geringsing wayang putri*. (IPCW)

Above: Ni Luh Kembang weaving double ikat, 2018.

Below: loom with double ikat from Tenganan Pegringsingan village, Bali. Ikat by I Made Katung, textile woven by Ni Luh Kembang. Wood from jackfruit tree, cotton, natural dyes.

Insular Southeast Asia

Above: *geringsing wayang putri*, Tenganan Pegringsingan village, Bali, woven by Ni Nyoman Diani, 2018. The warp runs horizontally in this photograph.

Below: detail of the loom on the facing page.

31 Sumba Body-Tensioned Loom for *Hinggi Kombu*

Gift of Sandra Sardjono.

This Sumba loom was acquired in Rende village in East Sumba in 2007. It is used for weaving a warp ikat textile for a man's cloth known as *hinggi kombu*. What is currently on the loom is only one half of a *hinggi*. A complete cloth would require two identical textiles, which would be seamed together along the selvedges. Notable features of this Sumba loom are the warp beam and the cloth beam. Here they consist of double wooden rods with finials in the shape of a horse. The materials used to make this cloth reflect locally available plants: cotton for the threads, red dye from the roots of *Morinda citrifolia*, and blue dye from the leaves of *Indigofera tinctoria*.

The bold patterns and deep colors characterizing this textile marks the high social status of the wearer. In the past, only people with noble status were allowed to wear such highly decorated textiles. Traditionally, they would be worn in pairs. The first is wrapped around the shoulders as a mantle and the second around the hips as a loincloth. But *hinggi kombu* is not only clothing for the living; it is also wrapper for the dead. As burial clothing, it protects the soul from malevolent forces, and its patterns and colors would identify the deceased to his ancestors. The motifs on this *hinggi kombu* are arranged in bands and show clever use of symmetries along both the vertical axis and the horizontal axis. The butterfly-like shapes on the blue ground mark the center of the cloth. The center section is often reserved for patterns derived from foreign cloth, signifying prestige. Above and below the center bands are rows of horses with double-headed birds, and deer with snakes. Horses and deer symbolize warfare, nobility and wealth. The end-bands with the red ground contain a row of cocks, which represent the daily life of the village. (SS)

Newly-made double ikat cloth, Tenganan Pegringsingan village, Bali, 2018.

Insular Southeast Asia

Sumba loom with partly-finished *hinggi kombu*, warp ikat, handspun cotton and natural dyes.
Loom width: 127 cm; textile width: 57 cm.

32 Simalungun Batak Body-Tensioned Loom for *Bulang* Textile

The Simalungun Batak weaving tradition is on the verge of extinction. The *bulang* textile, once worn as a headdress by women on all occasions, is currently worn only on a few ceremonial occasions. If production was once widespread, toward the end of the 20th century it was found only in the village of Nagori Tongah. Gradually, competition from (semi-) mechanized looms made it impossible for even this to continue. The loom and textile presented in this exhibition are part of a recent attempt to revive the *bulang* weaving tradition. Notably, the commission of this loom by the China National Silk Museum gave the effort a significant boost. The weaver, Ompu Elza, resumed weaving for the first time in 15 years; a significant challenge for herself. The tradition of loom making

Loom made by Obersius Damanik; textile woven by Ompu Elza, boru Sinaga, Nyonya Sitanggang. Wood from sugar palm, cotton, natural dyes. Loom width: 60 cm; textile width: 33 cm. Commissioned by the NSM in 2018.

Detail of warp patterning on the loom.

had also ceased, but her son-in-law made the loom through a process of trial and error and extensive consultation with weavers.

The *bulang* is one of the most complex Simalungun textiles with supplementary warp patterning in the red center field, and supplementary weft in the white end fields. The transition from red to white warp in the cloth is executed using a rare technique called warp substitution. When the red warp is half woven, the weaver "sows" in the new white warp, cuts out the unwoven red warp, and continues with the white warp. The red fringe and white loops of yarn at the join are tell-tale signs of the substitution technique.

The end fields are decorated with ancient supplementary weft patterning that involves the use of shed savers. In this loom, the weaver has finished half of the first motif. The shed savers allow her to finish the second half in mirror opposition. The patterning woven into one white end of the textile is never exactly the same as found at the other white end.

The textile is symmetrical along both horizontal and vertical axes, and also tripartite. Of all the textiles in the Simalungun repertory, this structure is most elaborated in the *bulang* and recapitulated in even the smallest technical details of weaving. A ritual cloth is about both process and product and cannot be produced on a (semi-) mechanized loom. (SN)

Ompu Elza weaving the bulang textile, 2018 (photo: Sandra Niessen).

33 Palembang *Songket* Loom

This loom was prepared in 2018 by the Zainal Songket workshop, and was made specifically for this exhibition. Palembang, the capital city of the Indonesian province of South Sumatra, is famous for the production of *songket* or supplementary-weft weaving with gilded threads. *Songket* is woven on a backstrap loom with two string heddles for the plain weave ground and multiple pattern heddles. The more elaborate the pattern, the more heddles are required. The loom displayed in the exhibition *A World of Looms* has hundreds of pattern heddles. *Songket* workshops continue to thrive in Palembang due to the patronage of upper-class society in Sumatra that appreciates the textile's high value and prestige associations. (ZA)

Palembang *songket* loom, South Sumatra. Wood, silk, gilded threads. See **section 11**, fig 11.1 to see the loom in operation.

Insular Southeast Asia

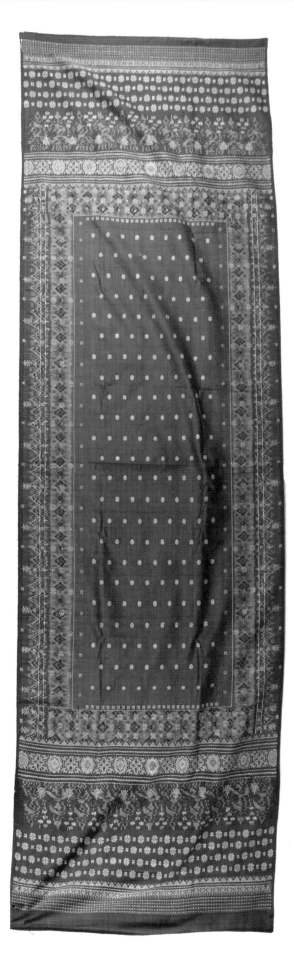

A pair of *songket* textiles, Palembang, South Sumatra.

Songket textile from Bali, woven by Ida Ayu Puniari.

South Asia

Introduction

The first loom shown here is an example of a pit loom, a basic type that is found across northern India as well as adjacent regions in Afghanistan and Pakistan. The two Indian looms that follow are examples of more complex types used for weaving luxurious cloths decorated with supplementary weft, including the Varanasi *jaala* drawloom that holds a permanent record of the pattern to be woven.

34 Bengal Pit Loom

The simplest and most widespread of Indian looms is the pit loom (*khadda shaal*) with treadles, mainly used for making plain woven yardage from fine handspun cotton (*khadi*), as seen in Bengal or southern India. The weaver sits on the edge of a pit dug in the ground and operates two or four treadles, which are linked to bidirectional (clasped) heddles in the warp. They are linked via heddle-

Pit loom.

A World of Looms: Weaving Technology and Textile Arts

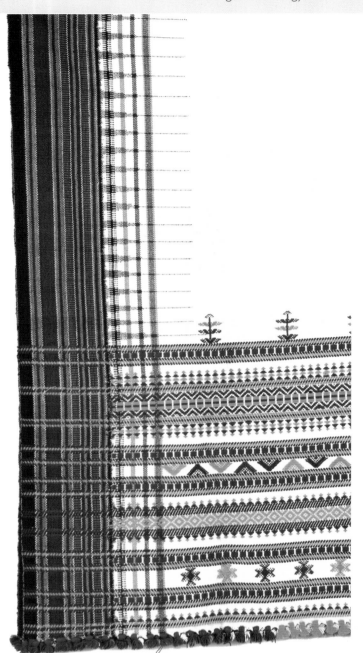

Textile woven on a pit loom.

35 Gujarat Multiple Heddle and Treadle Loom

In Gujarat, a modified fly-shuttle version of the pit loom (*khadda shaal*) is called *Ghoda shaal*. The loom shown here is a *Ghoda shaal* loom procured from Bhujodi, Bhuj, Gujarat.

This loom has twenty-one heddles linked to eleven treadles, and is used for making *kachchhi mashru*, a mixed-fiber textile in satin weave with silk warp and cotton weft. The pattern often consists of brilliant-color and contrasting solid and ikat stripes. The front part of a *mashru* would be warp-faced, exhibiting the brilliant silk satin finish, while the back—which would be worn against the skin—would be showing mostly the cotton wefts. (HA)

Gujarat multiple heddle and treadle loom.

horses above the warp, which are in turn suspended overhead from a frame or from the ceiling of the weaver's workshop. A reed in a heavy frame is used for beating-in the weft. There is a cloth beam in front of the weaver, but no warp beam as such: instead the warp is tied to a cord which is fixed to a stake in the ground. This can be loosened to release a new section of unwoven warp, as the woven part is wound on the cloth beam.

The simplest pit looms use throw-shuttles, but pit looms with fly-shuttles (a European import) are also regularly seen in India. (CB)

36 Varanasi *Jaala* Drawloom

Jaala drawloom.

A drawloom, in simplest terms, can be defined as a loom with two separate harnesses—one for the ground weave and the other for pattern—each running independently on the same warp. These looms are operated by skilled master weavers and their assistants. The most technically complex of Indian weaves are those that add weft patterning with the aid of a drawloom. This sophisticated mechanical device sits at the apex of traditional Indian silk-weaving technology.

The Varanasi *jaala* drawloom differs from other traditional Indian looms by the presence of a warp-beam (which is absent in most pit looms, such as cat. 34) and the absence of a counter-balance mechanism for the heddles. Instead, the drawloom has separate sets of treadles that control the lifting shafts and the depression shafts.

The key component of the drawloom mechanism is the *naqsha*, a detachable set of vertical draw-cords and horizontal pattern lashes, which acts

Detail of *jaala* patterning system. Vertical pattern leashes with embedded pattern cords descend and are attached to a set of horizontal cross-cords. These, in turn, are tied up in groups, representing pattern repeats, to the warp below.

as a template for pattern lifting. The *naqsha* is prepared on a frame called *machaan* and then installed at the upper level of the drawloom. The vertical drawcords of the *naqsha* are knotted onto horizontal cross-cords, which in turn hold groups of warp-ends with string leashes.

To operate the *jaala* drawloom, the drawperson pulls a single pattern lash of the *naqsha* to separate the draw-cords for that particular pattern pick and pulls these draw-cords by inserting and twisting a wooden fork (*mantha*). These pulled draw-cords lift their corresponding cross-cords and thereby the string leashes and the warp-ends connected to them, thus creating a patterning shed to pass a pattern weft. The weaver then inserts an angled-hook (*akda*) under the cross-cords to equalize the shed height across the loom width and also keep the shed open once the drawperson releases the pattern lash.

The number of warp-ends each cross-cord leash holds and the density of the warp-ends-per-inch determine the fineness and smoothness of pattern

Above: *machaan*, the frame on which the pattern master prepares the *naqsha* (web of pattern cords). Below: textiles woven on *jaala* drawlooms.

outlines and edges. These two factors, combined with the total number of cross-cords (which is equal to the total number of warp pattern-steps of the design), dictate the width of a pattern repeat. The configuration of the pattern harness does not in anyway restrict the lengthwise repeat of the pattern, which is controlled only by the number of pattern lashes in the *naqsha*.

The famous Indian gold- and silver-brocaded textiles included sashes, sarees and wrappers, along with fabric lengths for stitched garments and furnishings from the 16th to the 19th centuries. Some of these were woven in complex double-warp and pile-warp velvet techniques. All owed their magnificence to the exceptional skills of the Indian *jaala* drawloom weavers operating in the Mughal *karkhanas* of western and northern India, Varanasi, and a few other Indian weaving centers. While the products attributed to this era had some similarities to the Safavid-period drawloom textiles, the color palette, design principles, quality of metallic yarn, as well as the end products, were quite different. (HA)

Central and Southwest Asia

Introduction

Central Asian weaving traditions are closely related to those of northern India, and the looms used include some of the same basic types. These include ground looms fixed at four corners, pit looms and frame looms with bidirectional heddles. The *zilu* loom, a large frame loom, has some similarities with the Varanasi drawloom in the way in which pattern warps are controlled by a system of cross-cords suspended near to the warp, though the *zilu* loom lacks the pattern recording system of the Indian loom.

37 Ancient Egyptian Ground Loom

The use of the horizontal, two-bar ground loom can be traced back in ancient Egypt, via images on items such as the Badari bowl (section 14.2, fig. 14.1), to at least 3600 BCE, but its use is likely to be much older. There are also various Middle Egyptian (c. 2000 BCE) models and images in tombs of these looms, some of which represent workshop production of textiles (fig. 14.2). Parts of such looms (but no complete examples) have been found at various archaeological excavations. In addition, thousands of pieces of actual cloth have been recorded from archaeological sites and ancient burials all over the country. It is likely that this loom was closely related to modern ground-looms, such as the Bedouin loom (cat. 38).

The ancient Egyptian ground loom was normally made up of a warp and a cloth beam produced

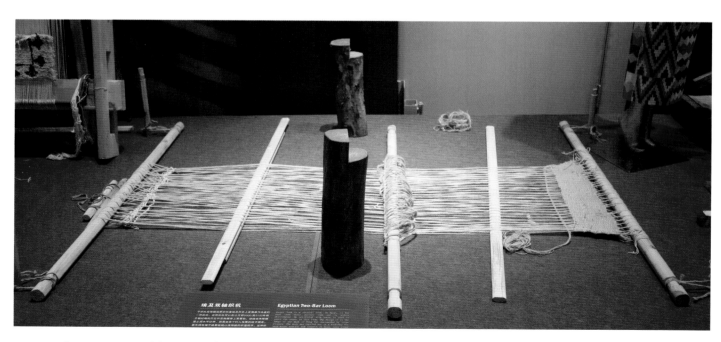

Reconstruction of the ancient Egyptian ground loom.

38 Bedouin Ground Loom

The modern Bedouin ground loom is a temporary construction: lightweight, easy and quick to set up and dismantle by nomads following their flocks of sheep and goats from one region to another. The loom is composed of a warp beam and a cloth beam, which are made from wooden poles. Each of these are kept in place with two pegs hammered into the ground. The size of the loom is dependent on the size of the finished cloth required. Some looms can have warps that are 15 m in length.

The warp is laid out on the loom by passing the yarn back and forth between the warp beam and the cloth beam, adding heddle loops to alternate warps at the same time. The warp is circular, but the cloth is woven as a flat sheet, uniting top and bottom parts as the weaving progresses. Different colored warps are often added to make striped textiles.

Like most other ground looms in North Africa and Central Asia, the heddle rod is supported above the warp on stones or blocks, and does not move during weaving. The default/resting shed available to the weaver is therefore the counter-shed. The natural shed is opened by pulling a large shed stick that sits behind the heddle forward, or (in some looms) by turning a flat shed stick through 90 degrees. This is sometimes done by the weaver herself, or sometimes by an assistant. The narrow sheds that are opened initially are widened by the weaver by inserting a sword-beater and turning it at right-angles, before inserting weft and beating it in. Beaters are made out of heavy lengths of shaped wood or, in some cases, camel ribs (these were particularly used for the weaving of narrow bands). Deer horns are sometimes used as "pin-beaters" to separate warps and to beat-in weft between warp yarns.

Bedouin ground loom weavers (who are usually women) normally sit on top of the woven portion of the cloth. As the work progresses the heddle is pushed down the long length of warp threads, with the warp and cloth beams remaining in situ, until the entire length of warp has been woven.

Weaving on the reconstructed loom.

from wood, each of which was kept in place with two pegs hammered into the ground. The beams were often tied to the pegs with ropes. The size of the loom was dependent on the dimensions of the cloth to be woven. Based on tomb models and paintings it would appear that they were usually about 2 to 4 m in length. Normally there was 1 (sometimes 2) heddles supported on jacks. The cloth was rolled onto the cloth beam as it was made. The loom seems to have been worked by two women, a weaver and an assistant. The weaver would be responsible for passing the weft thread through the shed and beating-in the weft using a heavy wooden beater. It was the job of the assistant to open shed and counter-shed, either by raising and lowering the heddle(s), or by moving a shed bar behind the main heddle to the left or right, as contemporary Bedouin weavers do. (GVE)

Most Bedouin looms produce two types of cloth: a plain woven cloth for garments such as long gowns (*thob* or a *dishdasha*) and outer garments (*abayah*), and a heavier cloth, either a warp-faced plain weave or a compound weave made from coarse goat hair, for making items such as tents, tent bands and tent dividers. The tent bands and dividers are often decorated with supplementary wefts in coloured wool thread. The designs tend to be geometric shapes, such as diamonds, triangles and waves. This fabric is also used for making saddle bags and girths for camels, donkeys and camels. These are often highly decorative and embellished with tassels. (GVE)

Bedouin loom.

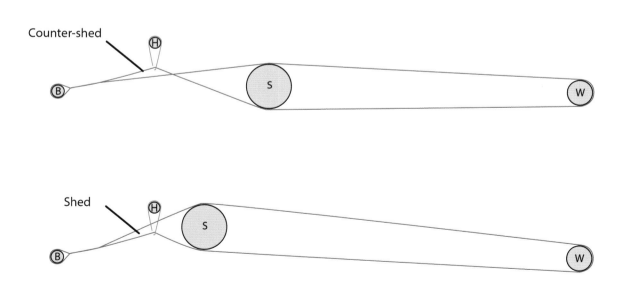

Diagram showing shed opening in a Central Asian ground loom. B—cloth beam, W—warp beam, H—heddle, S—shed rod. Top: with the shed rod (s) pushed back, the counter-shed is open. This is the default position. Below: with the shed rod pulled forward, the natural shed is opened in front of the heddle. The weaver inserts a sword-beater into these sheds to widen them for weft insertion (drawing: Christopher Buckley).

39 Egyptian Carpet Loom

Upright carpet loom.

The main type of loom associated with Egyptian carpets is the vertical loom, which varies in width from about 2 m to looms up to 10 m wide that would be used by several weavers sitting side-by-side. The basic carpet loom is made from a large upright frame with a warp beam at the top and a "carpet" beam (the same as the cloth beam on a textile loom) at the bottom. Sometimes the size of the frame is slightly larger than the required finished size of the carpet and the two beams are fixed in place. On other occasions the warp and carpet beams move so that the fresh warp threads are revealed and a much longer length of carpet can be produced. The warp threads are tensioned between the warp and carpet beam. Work putting in the individual pile knots (usually an asymmetrical type) can then commence. Each knot is wrapped around the warp threads by hand and then cut using a special knife. After each row of knots there are two or more rows of plain weave, which keeps the knots in place. In order to make a more compact weave a heavy comb is used to beat down both the knots and plain weave regions. In some cases, the design is worked on a piece of graph paper (cartoon) and placed behind the warp threads to be copied. On other occasions the design is worked from memory. (GVE)

Detail of partly woven carpet, with color yarns for piles hanging above, and graph to guide the placement of knots.

40 *Zilu* Loom

An Iranian *zilu* loom is a vertical loom. It consists of a frame made from two large upright posts (often roughly shaped tree trunks), both with a forked top. These two posts carry a cylindrical warp beam made out of a shaped tree trunk. At the bottom of the loom is the cylindrical cloth beam, which is also made out of a shaped tree trunk. The cloth beam is sometimes sited in a pit for extra space, as up to 10 m of thick woven cloth might be rolled around the beam, such as when a very large floor covering was made for the prayer room of a mosque. A *zilu* loom can vary in size from about 2 m to over 6 m in length, and about 3 m in height depending on the size of the textile being woven.

The warp threads are divided into two groups, namely the binding warps and the pattern warps. The binding warps are controlled by a pair of heddle rods, while the pattern warps are threaded through thick cotton leashes that are fixed to one

Zilu loom on display at NSM.

Detail of the patterning system, showing cross-cords and leashes, a wooden hook for holding open the pattern shed and a partly woven *zilu*.

Beating a weft yarn, *zilu* loom in Meybod, Iran, 2001 (photo: Gillian Vogelsang-Eastwood).

or two heavy cords (*maj*). The cords run between two vertical posts set just in front of the loom. The pattern warps can be raised individually or in groups, depending upon the design being produced. There is no system for pre-selecting the pattern repeat across the total width of the loom. Instead the weaver walks from one side of the loom to the other, raising and lowering the heddles and leashes as required for the design. The individual patterns, and the order in which the pattern warp threads are raised and lowered, are memorized by the weaver, rather than written down on a chart. The weaver is often assisted by an apprentice.

The weft threads are added by hand and are beaten-in with a heavy comb made of wood with metal teeth. Normally two colors are used for the weft, with blue and white being the most popular colors, although very occasionally red and white can be found. By the end of the 20th century, green, orange, and bright red had been added to the color repertoire.

This type of loom has been described as "a surviving representative of a pattern loom that was once much more widespread," and it is said that it may have been developed during the Roman Period for making fine silks as well as much coarser textiles.[10] If this is correct then it is part of a 2000-year-old weaving tradition that now appears to be coming to an end. (GVE)

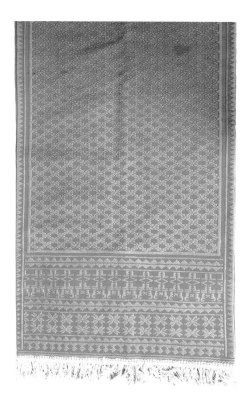

Zilu, gift of Gillian Vogelsang-Eastwood.

41 Margilan Ikat Loom

Gift of Rasuljon Mirzaahmedov.

This is the loom used for weaving cloth decorated with ikat in Central Asia. Since much of the intricate work of tying and dyeing has already been done by the time the cloth is woven (section 15), the loom is relatively simple. It is a four-post frame loom with a cloth beam fixed in front of the weaver, on which the finished cloth is wound. The warp passes over a beam at the other end of the loom and is tensioned with a weight. The loom is equipped with four heddles, linked together and suspended from pulleys. Four heddles can produce twills, but in this case the heddles are linked to just two treadles and are set up to make plain weave. A reed in a large swinging frame in front of the weaver is used to beat-in the weft. (CB)

Above: Margilan ikat loom, with ikat textiles behind. Facing page: ikat textile woven on the Margilan loom.

Africa

Introduction

From amongst the vast range of looms in the African continent, the looms selected for this catalog are frame looms used for weaving narrow strips of cloth in Ghana and adjacent regions, and a group of looms used in Madagascar that are related to ground looms used in Central Asia as well as other parts of Africa.

42 Ghana and Togo Loom (*Agbati*)

Many different kinds of handwoven textiles, commonly known as *kente*, are produced in the Asante region as well as the southern Volta region of Ghana and adjacent Togo. Each area use a different type of looms. These two types and the different textiles woven on them, with a wide variety of designs, have independent but interrelated histories that can be traced back to at least the 1600s.

Weaving centres in Ghana and other West African countries influence each other. In Southeastern Ghana and southern Togo, where Ewe is the first language of most people, exchange of techniques and design takes especially place with the Asante region and western Nigeria, where *aso oke* is woven. Today, the distinction Asante and Ewe weaving is gradually becoming blurred, though the distinction in looms endures.

Throughout this region, weavers use either six or four pole double-heddle looms. Some are fixed in the ground, others are mounted on a platform

Agbati (loom) from Agotime, Ghana.

Detail of warp- and weft-float weaving on the *agbati*, showing the shuttle and reed.

Gator Gbagbo, a well-known designer and weaver, in his loom in Agotime-Abenyiase, 2017 (photo: Malika Kraamer).

and therefore, in principle, portable. The oldest type is the six-pole loom and this was the only type until the end of the 19th century. This loom has four large poles of around 120 cm high that hold two top horizontal struts and support the warp beam; they are placed in such a way that they form a trapezium. Two poles of half the size, not connected to other parts of the framework, are placed in the ground in between and hold the cloth beam.

Out of this loom, the four-pole loom was developed at the end of the 1800s or early 1900s, when local carpenters, a trade introduced by the Bremen missionaries, constructed a portable version by mounting the framework on a trapezium, which required a reduction in the number of poles. According to local recollection, the use of a fixed four-pole loom developed only after the portable loom, which would explain why four-pole looms today, both fixed and portable, are only found in the town and larger villages, where there was enough work to support carpenters, but not in the hamlets. The invention of the four-pole loom happened roughly at the same time in the coastal and inland areas, including Agotime, but they developed differently from the same six-pole loom. The China National Silk Museum has a four-pole loom from the Agotime area in its collection (see photo on the opposite page, far left).

The loom is generally called *agbati*, meaning "loom or wooden frame," but the fixed and portable looms also have specific names, which differ in Agotime and the coastal area. Agotime weavers call the fixed loom *anyigbagba* (ground-loom) and the portable version *agbatata* (carpenter-made loom), while coastal weavers call them *tomegbati* (planted-loom) and *xormegbati* (loom-in-house) respectively.[11] The weaver sits on a stool behind the cloth beam. The warp extends through the beater and heddles, over the warp beam and is kept in tension by a weight on a sledge up to six meters in front of the loom. The sledge is pulled towards the loom every time the weaver rolls a strip of newly woven textile onto the cloth beam. This beam is prevented from unrolling by an iron or wooden pin placed through a hole on the right side of the cloth beam and braced against a stone on the ground. This pin can also be held in place by a wooden extension attached to the ground or to the loom, or by a cord connected to the top strut. The pin also functions as a lever to turn the cloth beam.

Weavers working on this loom, like weavers working on central Ghana loom, have explored almost every kind of weaving structure possible using one or two pairs of heddles. Ewe weavers are

A World of Looms: Weaving Technology and Textile Arts

Weft-faced plain weave cloth, locally called *titriku*, woven in 2018 by weavers in the workshop of Fred and Richard Agbo from Agotime-Abenyinase.

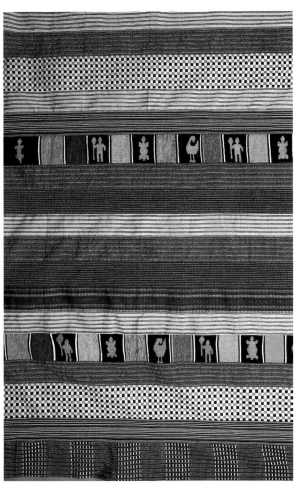

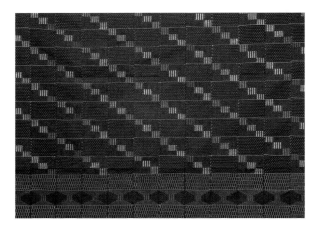

A supplementary warp and weft cloth, locally called *ebevor,* meaning "fishing-net-cloth," woven in Agotime-Kpetoe in 2017.

A cloth composed of strips in a variety of weave structures, locally called *sasã*, meaning "mixture," woven in Agotime-Kpetoe in 2017.

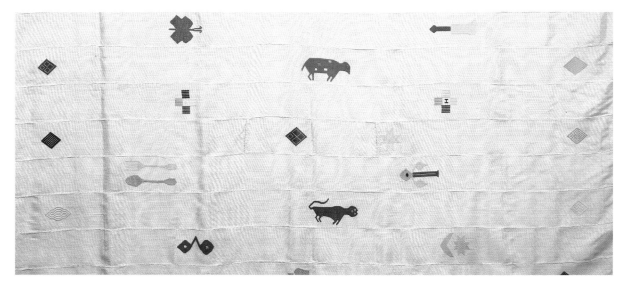

Asidanuvor (cloth with a handpicked design, only visible on one side of the cloth), woven by Edem Joshua Gally in 1999 in Agbozume, in the coastal area of southeastern Ghana, for the coming-out ceremony of the Agbozume Queen Mother.

also masters in creating different shades of color through intricate ways of laying the warp and the use of plied yarns of two colors. They make entirely warp-faced and weft-faced textiles, but through the use of two pairs of heddles, they are also able to alternate between weft-faced and warp-faced plain weaves in one length of woven strip. The use of supplementary weft to create designs, visible on both sides of the cloth, will be discussed in more detail in the section on central Ghana looms. In the past weavers in this region mainly made figurative designs in cotton, while today, like the much longer tradition in Asante, many also make cloth with non-figurative designs in either cotton or rayon. A design feature exclusive to this region is the use of a supplementary warp. Since the beginning of the 20th century, weavers have also begun to use single-heddles to create patterns visible on one side of the cloth. (MK)

43 Central Ghana Loom (*Nsadua*)

One of the principal courtly arts of the Asante Kingdom in central Ghana is the making of handwoven cloth in narrow strips, sewn edge to edge to form *kente* cloth, which emerged in its current-related form at the turn of the 18th century. The loom is this area, called *nsadua*, has four-poles placed in a rectangle, either portable or fixed in the ground. The loom shares features with looms to the north and west, while looms in the Ewe-speaking region have more similarities with those further to the east, such as those used in Nigeria. The fixing of the cloth beam, as well as the counting systems and method for mounting the warp are different in the several main weaving centers in Ghana. In Asante, weavers make predominantly warp-faced narrow strips with *asatia* heddles that are closest to the weaver, and

Nsadua Asante loom from Ghana.

Detail of *Nsadua* Asante loom, showing a shuttle, a reed beater and three pairs of heddles that are used for patterning the woven cloth, based on twill weave.

an alternation of weft-faced and warp-faced plain weave blocks by using a second pair of heddles, *asanan*. After laying the warp, the weaver must first pass the warp elements in units of six through the *asatia* heddles and then one by one through the *asanan* heddles to achieve this alternation, in the same way as weavers do in the predominantly Ewe-speaking region. The *asanan* heddles are not just used to create weft-faced plain weave bands, but also to weave supplementary-weft float patterns. One of the most complicated techniques employed on the central Ghana loom involves the use of three pairs of heddles, one for the ground weft and two for the creation of supplementary-weft floats. The extra pair of heddles allows for an overlapping diagonal arrangement of floats characteristic of twill weaves.

Kente are the most famous cloths from West Africa, and as such their distinctive designs have been reproduced in large quantities on printed cloths, particularly in factories in China, for export to Ghanaian and other African and American markets. (MK)

Facing page: cloth woven by Opanin Osei Kwame at the end of the 1990s for Nana Amma Serwaa, Queen Mother of Bonwire, Asante, Ghana. The cloth shows an alternation warp- and weft-faced plain weave with supplementary weft-float motifs. This cloth is a very large female cloth (the size sits between a female and male cloth). It is only used by the mothers of queens, worn over the shoulder.

Kwame Kusi Boateng weaving in an Asante loom. Bonwire, Asante region, Ghana, 2018 (photo: Malika Kraamer).

44 Madagascan Loom for *Akotifahana* Textile

The loom for *akotifahana*, Madagascar's historic brocade, is found throughout the central regions in the areas where mulberry silk is produced, such as Antananarivo, Antsirabe, and Arivonimamo. It consists of a simple frame on legs that allows seated weaving. Because of the fragility of the silk yarn, which is lightly twisted to maximize its luster, the pieces of cloth that are produced are relatively short with a continuous warp, yielding cloths around 250 cm long by 80 cm wide. The row of string heddles called *haraka* allows for the lifting of the warp threads to create the plain weave ground. The row of string heddles is created anew each time the weaver prepares a new warp. Sometimes it is prepared during the warping and sometimes it is done after the warp has been mounted on the loom.

The loom is equipped with extra heddles for supplementary weft patterns. These heddles are made of raphia or sisal fibers, sometimes also of synthetic fibers. The heddles are long enough to allow warp movement. They are tied on a small cane and organized by groups so that the weaver can easily choose them according to the pattern he is weaving. Each weaver has his own copybook where he notes all the patterns and the way to weave them according to the rhythms of lifting the warps. (AE)

Loom for *Akotifahana* textiles.

A World of Looms: Weaving Technology and Textile Arts

Loom for *akhotifahana* textiles and detail of brocade patterning.

45 Madagascan Loom for *Akotso* Textile

The loom for *akotso* textile is similar to that for *akotifahana*. The loom includes only one row of string heddles for the plain ground weave, and no extra heddles for the motifs. The weaver instead creates the patterns from memory. The supplementary wefts are often in pure white cotton. *Akotso* means "border" and is so-named because of the large white borders. This fabric is mainly intended for funerary use and rarely used for clothing. Akotso comes from the region of Ambalavao, in the south of the island. One may compare the technique used for the extra wefts to "embroidery in the warp threads," which can give complex and wonderful patterns, though the patterns produced have tended to become simpler in recent times. (AE)

46 Madagascan Loom for Raphia Textile

This loom is used for weaving textiles from *rabane* (raphia), *jabo* (raphia and silk or raphia and cotton), and *lay masaka* (ikats of raphia, or raphia and silk, or cotton), on the northwest coast of Madagascar between Mahajanga and Besalampy, but it can also be used for other types of fabric.

It is the simplest type of loom, consisting of two wooden beams that are placed on the ground and that are not connected to each other. The warp circulates around these beams. The length of the beams is around 3 m to 4 m. Wooden stakes hold the front and the back beams, with ropes at the front to maintain the tension in the warp. A row of string heddles is suspended between two wooden supports placed on each side of the loom. On the northwest coast, this row of heddles is always prepared during the warping, and installed on the loom around a wood cane and a little iron bar. It is suspended so that the weaver can move it a little along the length of the warp to facilitate weaving.

Africa

Loom for raphia textiles

Detail of ikat textile on the loom.

273

Raphia textile with indigo dyeing and ikat decoration.

From time to time the warp is relaxed and turned a little to continue the weaving.

Because they derive from palm tree leaves, raphia yarns are short and joined each meter, so they lack flexibility and are prone to breakage during weaving. This characteristic affects the length of the loom. For raphia ikat textiles called *lay masaka*, the loom is used for both the tying process before dyeing and the weaving. First, weaver(s) install the warp is on the loom, and then in the horizontal position they begin to tie the ikat pattern. When the tying process is finished, the warp is taken off the loom and brought to the dyer who will dye it with one, two, or three colors, sometimes more. After the dyeing process, the warp is returned to the loom. The weaver aligns the yarns carefully and then begins to weave. (AE)

Europe

Introduction

European weaving traditions are a tale of gradual transformation: an early type, the warp-weighted loom, was gradually replaced by horizontal frame looms and upright looms with two bars. Two looms that illustrate these types are shown here: a reconstruction of an ancient warp-weighted loom and a horizontal loom used for making fine tapestry.

Reconstruction of a Roman Period warp-weighted loom.

47 Ancient Warp-Weighted Loom

A warp-weighted loom stands upright or at an angle, leaning against a wall or a ceiling beam. The vertically hung warp threads are kept taut by weights attached at the warp ends. During weaving, the weaver stands in front of the loom and weaves from the top downwards, beating the weft upwards. The number of sheds required for particular weave determines the number of heddle bars needed. A warp-weighted loom can be used to produce many different types of textiles, but it is especially suited for weaving twills.

Archaeological and historical visual evidence show that loom weights vary in materials, shapes, thickness, and weights: some are made of clay or stone, some are round, pyramidal or conical in shape, and some are thicker and heavier than others. These variations reflect certain regional and temporal specificities. The loom in the exhibition *A World of Looms* was a reconstruction of a type that was common throughout the Mediterranean and especially widespread throughout the Roman world. It has pyramidal loom weights that were modelled after the archaeological finds from Insula VI.I in Pompeii.[12] The heaviness and thickness of the weights make the loom suitable for weaving slightly coarser textiles. The exhibited loom also matches the comparatively slender loom frames seen in Mediterranean depictions of warp-weighted looms.

In order to attach the warp threads to a loom weight, a loop is tied through the loom weight hole and the warp threads are attached to the loop. When using stone or clay weights without a hole, a loop is tied around the weight. To weave a plain weave on a warp-weighted loom, alternate warp threads are placed in front of or behind a rod or shed bar on the loom. Two rows of loom weights

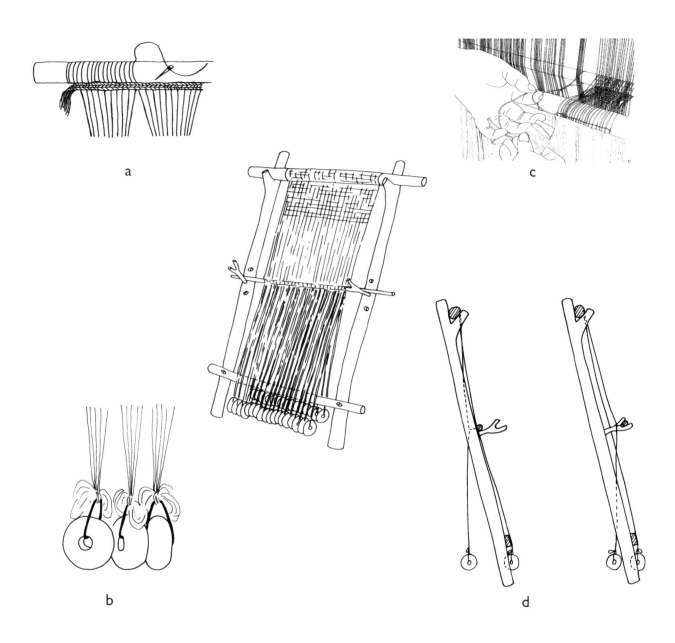

Setup for a warp-weighted loom making plain weave: a: tying the starting border; b: heddling; c: fastening the loom weights; d: changing sheds (drawing: Annika Jeppsson © Annika Jeppsson and Centre for Textile Research Copenhagen University).

are generally used. One row of loom weights hangs in front of the shed bar, the others hang behind it. The first warp thread is fastened to a loom weight in the front layer, the second to a loom weight in the back layer, the third to the front layer and so on. Depending on the size of the loom weight, a variable number of threads are attached to each individual weight.[13] (MÖ and EAS)

48 Aubusson Tapestry Loom

European tapestry may be made on a two types of looms, either a vertical loom (*haute-lisse*) or a narrow horizontal loom (*basse-lisse*) such as the one shown below from 18th century France. Both types co-existed in France and Flanders.

A tapestry is characterized by having discontinuous ground wefts that also act as pattern wefts. Multicolored wefts are worked into the warps in sections rather than in one pass across the selvedges. Each colored weft is hold in a shuttle-like device called a *broche* or *flute*. Today there are many weavings that claim to be handmade tapestry. But a true *tapisserie d'Aubusson* can be identified by looking at the reverse of the textile and searching for evidence of "wrong-side weaving" in the form of network of colored threads, which are the weft ends. The same is true for handmade tapestry produced in the tradition of Beauvais, Arras, Angers, the Gobelins, or Flanders. All of these were worked with the reverse side up facing the weavers.

A *tapisserie d'Aubusson* weaver usually weaves about 1 m^2 in a month. The workshops at Aubusson use the horizontal loom (*basse-lisse*). The loom is made of two-sided frames (*jumelles*). The frames support two large wooden cylinders that act as the warp and cloth beams (*ensouples*). The warp is rolled on one of the beams. The other beam holds the woven cloth.

At an early stage of the weaving set-up, a cartoon is placed under the warp. Typically, the cartoon painter and the weaver consult with each other on

Aubusson tapestry loom with European wall tapestries in the background.

the color scheme. The weaver first rolls the skeins onto bobbins and then fills up the shuttles in process called *flutage*. The weaver may put several weft threads (usually wool) an average of four or five, in a shuttle, to make either monochrome or polychrome effects. The weaver then sets up the weft threads with a metal scraper (*grattoir*) and beats them in with a boxwood reed (*peigne*). A bodkin (*poincon*) is used to assist with the leveling of the warp when the tapestry is rolled. To check the quality of the finished work, the weaver uses a mirror.

To weave the first passage (*passee*) the weaver presses on one treadle, which lowers the even warp, and with the right hand slides the shuttle loaded with wool inside the warp shed towards the left hand. When the other treadle is pressed, the weaver lowers the odd warp and slips the shuttle from left to right. After several passages, the wefts are lightly compacted with a comb. To create shading effects, the weaver employs hatching and gradation in colors (*battages*), shifting from one hue to another. (AM)

Detail of the Aubusson loom.

The Americas

Introduction

Two looms are shown here that illustrate the variety and versatility of weaving on body-tensioned looms. The looms are from Peru, but weaving on similar looms is found along much of the eastern seaboard of the Americas, from Mexico through Ecuador and into Peru.

49 Chinchero Body-Tensioned Looms

Chinchero is a small town in the Urubamba province outside of Cuzco, in the mountain area at 3,763 m above sea level. It is known today for its active Sunday market and especially for its expert weavers. The looms in the exhibition come from Chinchero and display blankets in the process of weaving. Weaver Flora Callanaupa is shown using one of these looms in section 23.

The first loom, a backstrap loom with warp-patterned weave, displays the components of a traditional Andean loom: the wooden loom bars, notched at each end. On one end a belt would be attached and tied around the weaver's waist. On the other, there would be a cord that is tied to a fixed pole or tree to secure and tension the warp. For making plain weave, one shed bar maintains the natural shed, and a set of heddles maintains the counter shed. The heddles are created after the warp is on the loom and attached to a small rod. Additional shed sticks may aid the weaver in creating the patterns, though these are usually

Chinchero loom.

created individually, by hand-picking the warps to form a complementary two-faced weave.

The partially woven textile is half of a woman's shoulder mantle with warp-faced and warp-patterned weaving. The patterning is formed in the long, narrow vertical stripes composed of different color yarns (often laid out in pairs) that are placed on the loom during the warping process. The weaver creates the pattern from memory, sometimes using an existing mantle as her guide. The sharp bone tool is used the pack the weft yarns within the densely set warp. A finished mantle is woven in two parts, sewn together down the center. Two loom bars, a shed rod, a heddle rod, and 3 pattern sticks separate the pairs of colors used for warp patterning. The weft yarn is wound on the long stick for insertion.

The second loom has a partially-woven textile with warp-faced discontinuous-warp plain weave and warp patterning. The weaving is created with four-color changes in the warp. These color changes are planned during the warping process, which requires additional sets of scaffolding sticks for each junction of the warps, which are dovetailed around the sticks. During the weaving the sticks are changed to heavy cord. The process is very time-consuming and requires a great deal of thought and skill; each section contains warp-patterning areas, whose colors and patterns change from section to section. To make the densely-set warp-faced weaving, additional sets of heddles and pattern sticks need to be set up for each section separately. The weaver proceeds through each section in turn: a *tour-de-force* of mental, physical and artistic effort. (EP)

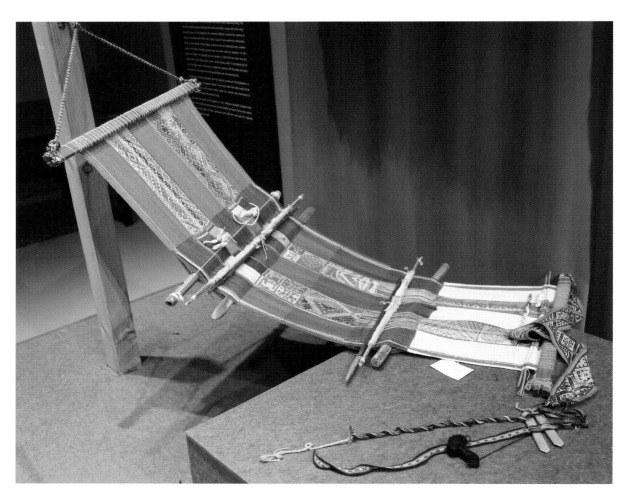

Chinchero loom set up for warp patterning on a discontinuous warp.

The Jacquard Loom and After

Introduction

The story of modern weaving is one of adding increasingly sophisticated control mechanisms to the weaving process. Two looms are described here: one that represents the beginning of this mechanical revolution, The Jacquard loom, and one that represents the current "state of the art" in computer-controlled weaving, which continues to advance to even greater levels of control and precision.

50 Jacquard Loom

Until the beginning of the 19th century in Europe, the drawloom was the main loom for weaving patterned textiles. This loom has several shafts controlled by treadles to weave the ground, while the design is controlled by the simple-cords and the pattern lashes, operated by the drawperson.

The "Jacquard machine," in contrast, is a card-reading device. This machine is the result of a century of research that started in 1725 by

"Jacquard" device for controlling warp lifts on a loom, using a punched-card system.

a Frenchman from Lyon, Basile Bouchon. As noted in section 24, this machine took its name from Joseph Marie Jacquard who nevertheless contributed relatively little to its development and held no patents on the device; the only patents being originated by Breton. Despite its early beginnings, it was not until 1817 that the Jacquard loom ran efficiently.

The Jacquard loom works using punched cards on which each hole is matched with one warp yarn. For instance, if position 35 on the card is punched with a hole, the yarn number 35 will lift up; if the spot is not punched, the yarn remains stationary.

In term of the mechanics of the Jacquard machine, the holes in the cards control horizontal pins, and the movements of these pins control some vertical hooks that are tied to necking cords and heddles. If there is a hole, a pin goes through and the hook remains stationary. In this position, the hook is caught by a rack system and lifted up. Consequently, the corresponding necking cord, heddles, and yarn are raised. If the pin hits a card position with no hole, the card pushes the pin, and the pin pushes the hook. As a result, the rack does not catch the dislocated hook, and the heddle doesn't move. One card is used for each weft shot, so the number of cards corresponds to the length and the fineness of the design.

The 19th century "traditional Jacquard loom" from Lyon is actually equipped with two Jacquard devices: a small one with 104 hooks to control only the ground weave of the fabric, plus a bigger one with 400 or 800 hooks to control the design. The weaver works alone and controls these two machines with one or two treadles: one treadle for damask fabrics and two treadles for brocade or velvet fabrics. (GS)

51 Computer-Programmable Loom

The Jacquard loom was introduced into China via Japan at the end of the Qing Dynasty and the early years of the Republic of China. The Japanese had improved it before the loom was brought to Jiangsu province and Zhejiang province in 1910. The loom was equipped with a Jacquard shedding device and hand-pulled cords for controlling the movement of the sheds. Around the city of Suzhou, this loom was known as *tieji*, or "iron machine."

In 1915, the Shanghai Zhaoxin silk factory started by Shen Huaqing introduced nine Swiss-made electric looms for weaving silk; these were the first automatic looms in China. By the 1950s, many types of electric looms were in use in China, such as rapier loom, air-jet loom, gripper loom, and water-jet loom. With the disappearance of the traditional wefting (weft-insertion) using shuttle, production output increased exponentially. By the 1970s, with the advent of the multi-phase loom, continuous wefting began to replace intermittent wefting.

At present, there are nearly 2 million looms in the weaving industry in China. The types of looms commonly used include rapier loom, water-jet loom, air-jet loom, projectile loom, and shuttle loom. Of these, the number of the shuttle-less looms is now around 900,000 and accounts for 45% of all the industrial looms. The annual output of shuttle-less looms is approximately 70 billion meters.

Looms continue to develop, and the current general trend moves towards electromechanical integration and computer-assisted weaving. Energy-saving and environmental protection also become increasingly important considerations. (LJ)

The Jacquard Loom and After

Contemporary loom made by the Panter company in Italy. This is an example of a modern high-speed loom in which the warps are lifted by electromechanical actuators rather than the leashes employed in older looms. The loom is controlled by computer, and it can weave complex weave and designs that rival photography in their level of detail and color range.

Catalog Notes

1 Zhejiang Provincial Cultural Relics and Archaeology Research Institute 1988, 1–31.
2 Zhao 1992, 108–111.
3 Zhao 2012.
4 Boudot and Buckley 2015, 33.
5 The results of the project were published in Zhao et al. 2017.
6 Zhao 1997.
7 Boudot and Buckley 2015, 35–36.
8 Zhao 1994.
9 Zhao 1996.
10 Thompson and Granger-Taylor 1995–1996, 27.
11 In the 19th century, the word *agbati* was recorded by the Bremen missionary Spieth in the Ho area (Spieth 1906, 407). *Agbati* was probably the only term for loom until the 20th century, when portable looms were introduced next to fixed looms and a differentiation was made between the two, also in local terminology.
12 Baxter, Cool, and Anderson 2010.
13 Andersson Strand, Nosch, and Olofsson 2015.

Glossary

Abaca: (*Musa textilis*) a type of banana used as a source of *leaf fiber* for weaving cloth (amongst other uses), particularly in the Philippines.

Aniline dye: the earliest type of synthetic *dye*, made from aniline (a coal-tar extract). Discovered in 1856, aniline dyes were commercialized for a short period in the 19th century, but had very poor light fastness and were soon replaced by other types of synthetic dye. The terms "aniline dye" and "chemical dye" are sometimes used (incorrectly) to refer to synthetic dyes.

Backstrap: A strap around the weaver's waist, used for tensioning *body-tensioned looms*.

Backstrap loom: see *body-tensioned loom*.

Balanced weave: a woven surface on which the *warp* and *weft* are equally prominent.

Bamboo basket: a component of some *complex pattern heddles*. The basket takes the form of a cylinder made of bamboo strips, with a diameter varying between 40 cm and 60 cm, around which *pattern sticks* are held in sequence. Also called a "pig basket."

Bar heddle: a bar with grooves in it through which some *warp* yarns pass. When the bar is rotated, the warps in the grooves are lifted and lowered, creating sheds for the insertion of wefts. This type of *heddle* is used on some looms for making mats in the East Asia region.

Bark cloth: see *bast*.

Basket weave: *plain weave* based on groups of warp and weft yarns, instead of individual yarns. For example two or three warps/wefts are interlaced as if they were single warps/wefts.

Bast: the soft "inner bark" or phloem of some plants (such as hemp, ramie, jute, nettles, and flax), from which fibers can be extracted for making yarn. Bast can also be used to make non-woven cloth (*bark cloth*) by softening it through soaking and beating it out into a flat sheet.

Beater: a wooden stick (sword) or a *reed*, used to beating weft during weaving.

Bidirectional heddle: a heddle that has strings attached both above and below the warp, which can pull warps both upwards and downwards. Also called a clasped heddle.

Bobbin: see *shuttle*.

Body-tensioned loom: a loom that is tensioned by the weaver's body, generally via a backstrap around the weaver's waist. Body-tensioned looms include both frame and non-frame types. Synonyms: back-tensioned loom, backstrap loom, loin loom (India).

Brocade: In strict technical weaving terms, brocade refers to *discontinuous supplementary-weft* patterning. The term is also used more loosely by some authors for cloth with multi-colored designs that are added during weaving.

Calendering: a finishing process by which a smooth surface is produced on cloth. Various surface treatments may be used, including applying coating materials and rolling or hammering. Also called polishing or glazing.

Camelid hair: hair obtained from camelid species, including camels, llamas and alpacas.

Cantilever frame: a type of *frame loom* characterized by *heddles* and/or *reed* that are supported by a cantilever structure, attached to posts at the back of the loom.

Card weaving: a technique for making narrow bands and the starting borders for some types of weaving, using a pack of plates or cards with holes in. Warps pass through selected holes. The cards are rotated to move the warps up and down, opening and closing sheds for weft insertions. Card looms are usually body-tensioned, with a *cloth beam* attached to the weaver's waist. Also known as tablet weaving.

Carpet: a textile designed as a floor covering. Carpets may use knotting techniques to create *pile* or they may be "flatwoven" using techniques such as *tapestry*. Synonym: rug.

Chaîne opératoire (French): a fancy word for "process," such as the process for making a textile.

Circular warp: a warp that is mounted on the loom as a continuous loop, and that results in a textile that is in tubular form (usually with a short section of unwoven warp).

Clasped heddle: see *bidirectional heddle*.

Cloth beam: the axle/beam to which the warp is fastened (or passes around), and on which the woven fabric is wound on some looms. On horizontal looms this beam is usually positioned directly in front of the weaver, in which case it may also be called a breast-beam.

Coil rod: a rod around which every warp thread is wrapped once. This rod serves to keep the warp threads aligned and in order. Mainly found in simple *body-tensioned looms* that lack a *reed*.

Comb: see *reed*.

Complementary warp or weft: a group of warps or wefts that belong to a set; they are part of the interlacing for the *ground weave* (thus they are not *supplementary*) and also used to create patterns. Weave structures with complementary elements can be *warp-faced* or *weft-faced*. See *samit*, *taqueté*, and *jin*.

Complex pattern heddle: mechanism consisting of heddle strings with multiple *pattern rods or strings/cords* embedded in it, used to raise selected warps in sequence. Also called compound pattern leash system and *drawloom* system.

Compound pattern leash: see *complex pattern heddle*.

Compound weave: a *weave structure* with multiple sets of elements (*warp* and *weft*), each set fulfilling different roles, either for *ground weave* or for patterning.

Continuous supplementary weft: *supplementary weft* that extends from one edge of a woven fabric to the other edge (*selvedge*).

Cotton: A shrub plant from the *Gossypium* genus, whose seed heads produce fine floss fibers from which cotton *yarn* can be spun.

Counter shed: one of two sheds used in making *plain weave*, the other being the *natural shed*. The counter shed is usually opened by pulling up a lower layer of warp using a *heddle*.

Damask: A general term for monochrome patterned fabric with contrasting textures. These textures reflect the light in a different way, making the pattern visible. The technical definition of damask, used in this book, is a *simple weave* that combines—on a single layer—two types of surfaces, usually one would be *warp-faced* and the other *weft-faced*. In a strict technical term, damask applies to the combination of the two faces of the same weave, for example the warp-faced and the weft faced weaves of a *satin*. When the combination is composed of different weaves, such as plain weave and twill, the structure is called damassé.

Diamond twill: see *twill*.

Discontinuous supplementary weft: *supplementary weft* that extends across just a few warps, i.e. does not extend from selvedge to selvedge.

Discontinuous warp: a warp that is not wound continuously on the loom. Typically the ends of the warp are tied individually to the warp beam and the cloth beam, for example warps made of individual lengths of *raphia*.

Dobby: a Jacquard device for controlling a pattern loom.

Dovetailed tapestry: a type of *tapestry* in which the wefts of adjacent color sections share a common terminal warp.

Double cloth: a *compound weave* employing two sets of warps and wefts, each set interlaces to form its own weave and create a distinct layer, thus the name double cloth. Normally the two weaves are of the same type, for example, plain weave double cloth.

Double-heddle loom: a loom that has a pair of heddles for opening the sheds for making the *ground weave*. Other heddles, such as *pattern heddles*, may be present in addition. Compare *single-heddle loom*.

Double ikat: see *ikat*.

Draw-cord or draw-string: a leash/ *heddle string* in a *complex pattern heddle*.

Drawloom: a loom with a *complex pattern heddle* that records warp lifts for weft insertion, and allows warp threads to be raised in any combination. In addition to the weaver, there is normally a second person (drawperson) who manipulates the *pattern heddle*.

Dye or **dyestuff:** a soluble coloring material that penetrates and binds to fibers (as opposed to a *pigment*).

Fiber: the base material from which thread (yarn) is twisted, knotted, or spun.

Figure tower: see *drawloom*.

Flat warp: a warp that is not a *circular warp*.

Flat weave: see *carpet*.

Flax: see *bast* and *linen*.

Flying shuttle/fly shuttle: a shuttle that is propelled through a *shed* by a mechanism that "shoots" it through the warp opening, by pulling a cord or other actuating device. Flying shuttles were invented in Europe but are now found on frame looms worldwide.

Four-selvedged weaving: a type of South American weaving in which both the weft and warp edges loop around and form continuous, uncut selvedges, as opposed to two-selvedged cloths (with selvedges on the weft borders only) produced by most weavers.

Frame loom: a loom with a rigid frame around it.

Gauze: any sheer textile with an open structure is often called gauze, but in the sense used in this book, gauze refers to a specific type of weave that is made using crossed warps. Typically, gauze is derived from *plain weave*. The warp crossings create small openings in the ground weave, and patterns can be made by juxtaposing the ground weave and the gauze areas.

Ground weave: the warps and wefts that make up the basic structure of a textile, considered in isolation from any patterning elements that may be present. The ground weave may be a *plain weave*, *twill*, or a *satin* structure. Synonym: foundation weave.

Ground weave heddle: a heddle that raises or lowers the warps needed to make the *ground weave* of a textile.

Foundation weave: see *ground weave*.

Half-frame loom: a loom in which the *warp beam* is fixed in a frame, but the *cloth beam* "floats" and is attached to the weaver's waist.

Harness: a *leash*, or a set of leashes attached to warps on a pattern loom.

Heddle: A set of string loops (leashes) attached to warps, used for opening a shed for the insertion of weft. A simple heddle consists of a stick with loops of thread that can open one type of shed. *Complex pattern heddles* used on drawlooms can encode many shed openings.

Heddle string/loop: a single loop of a heddle, attached to a warp and used for lifting it up for weft insertion.

Hemp: (*Cannabis sativa*) plant from which hemp fiber can be extracted for making textiles.

Ikat: a resist-dyed process for making designs on yarns before weaving. The design is reserved by tying off small bundle yarns with strips of palm leaf (or plastics nowadays) to prevent the dyes from penetrating the fibers. This can be done on the warps (warp ikat), wefts (weft ikat), or both warps and wefts (double ikat). In most cases warp ikat is a *warp-faced weave*, weft ikat is a *weft-faced weave* and double ikat is a *balanced weave*.

Interlocked/interlocking tapestry: a type of *tapestry* in which the wefts of adjacent color sections loop around each other when they go back and forth.

Jin or **jin brocade:** Chinese term for ancient polychrome textiles, usually in *warp-faced compound plain weave*.

Kelim (kilim): a flatwoven (non-pile) floor covering. The technique of kelim is *slit tapestry*.

Kesi: Chinese term, literally translates as "cut silk," used to refer to woven patterned silk in *slit tapestry* technique.

Kudzu or ko-hemp: (*Pueraria sp*) a type of *bast* fiber made from the wild kudzu vine.

Lampas: a compound weave structure employing two sets of warps and wefts. Each set interlaces to form its own weave, but does not create a distinct layer as in *double cloth*. One set forms the *ground weave*, the other the pattern. Lampas textiles display contrasting weave surfaces, most often with the ground weave being *warp-faced* and the weave for the pattern *weft-faced*.

Leaf fiber: fibers extracted from leaves, such as banana, *abaca* and *raphia*. Compare *bast*.

Lease rod: a rod, sometimes a pair of rods, usually inserted during warping and used to maintain a warp crossing. Generally found close to the warp beam.

Leash: see *heddle string*.

Linen: a *bast* fiber obtained from flax (*Linum usitatissimim*).

Loin loom: see *body-tensioned loom*.

Loop-heddle: a type of *heddle* that consists of loops of string (*leashes*) that enclose warps.

Multi-heddle patterning system: a system in which warp lifts for patterning are recorded on multiple simple heddles, each heddle recording the warp lifts for one weft insertion (compare *complex pattern heddle*).

Natural shed: one of two sheds used for making plain weave fabrics, the other opening being called the *counter shed*. On simple looms the natural shed is held open with a stick (*shed stick* or *shed rod*).

Necking cord: part of a *drawloom* or a Jacquard loom; a cord that connects a *leash* to the patterning device.

Pattern cord: a cord that records a warp lift in a *complex pattern heddle*, such as that found in a *drawloom*. This term is also used by some authors as a synonym for *leash*, when referring to *drawlooms* and Jacquard looms.

Pattern/patterning heddle: a *heddle* that raises groups of warps for patterning purposes, for example for inserting *supplementary wefts*.

Pattern rod/stick:

1. Stick inserted directly into the warp, to save a shed opening for weft insertion. Generally used to save a *pattern shed* on a temporary basis.

2. Stick, usually made of bamboo, fixed permanently in a *complex pattern heddle* system, used for saving a series of shed openings for patterning purposes. This may be replaced by a cord (*pattern cord*) in some looms.

Both types of stick are called "pattern rod" or "pattern stick," though they are fundamentally different in terms of placement and usage.

Pattern shed: an opening in a warp created for the purpose of inserting patterning weft(s).

Patterning tower: a characteristic feature of a Chinese *drawloom*; a framework built above the warp that supports the patterning system and the drawperson who operates it.

Pedal: a treadle worked by the weaver, usually used to lift a *heddle*. The ensemble of treadle plus heddle is called a *shaft*.

Picking (warps): the process of selecting individual warps by hand, usually done with a stick or a hook. One "pick" refers to a group of warps that is raised by a weaver prior to inserting weft; it may also mean one weft insertion.

Pick-up stick: a pointed stick used by a weaver for selecting individual warps, prior to inserting a pattern weft

Pigment: a coloring material consisting of insoluble particles, for example mineral colors. Compare *dye*.

Pile: elements that are designed to protrude above the surface of the textile, such as the weft "knots" added to some *carpets* and the *supplementary-warp* loops of *velvets*. Pile elements are generally added during weaving, but may also be added after taking the textile off the loom.

Plain weave: the simplest weave structure, in which wefts pass over and under alternate warps. Sometimes called "tabby" or "plain tabby" (these terms are not used in this book).

Plate heddle: a plate with holes and slots in, through which warps are passed. When the plate is raised or lowered the warps in the holes are also raised and lowered, while the warps in the slots remain in position. By this means sheds are opened in the warp for insertion of wefts. This type of heddle is usually used for weaving narrow bands of cloth.

Plied yarn: yarn consisting of several strands, normally, but not always, twisted together.

Punched-card: a card with holes in, used for controlling movements of warps (such as in a Jacquard loom). Each hole corresponds to one warp lift: if a hole is present, the warp is raised and the weft passes underneath it during weaving; otherwise, the warp remains stationary.

Ramie: (*Boehmeria nivea*) a plant from the nettle family used as a source of *bast* fiber for weaving.

Raphia: (*Raphia sp*) a type of palm used as a source of *leaf fiber* for textile manufacture in some parts of Africa.

Reed: a device resembling a comb, used for keeping warp threads in sequence and evenly spaced. Usually positioned near the weaver, just in front of the leading edge of the fabric being woven. In some looms the reed is also used to beat-in the weft.

Rug: synonym for *carpet*.

Samit: a weft-faced compound twill, with two sets of warps (called ground/foundation and inner warps) and two or more *complementary wefts* (usually in different colors). All wefts and the foundation warp interlace in twill weave. The inner warp separates the wefts: one color to appear on the front of the fabric for making the pattern, and the rest to remain on the back. Compare *taqueté*.

Satin: a weave structure having warp- or weft-floats with interrupted diagonal bindings. It is also a general term (not used in this book) for fabrics with silky and glossy appearance. Compare *twill*.

Selvedge (selvage): the long edges of a woven textile that run parallel to the warp, created by turning the weft at the edge.

Shaft or harness: this term is used variously in textile literature to mean a *heddle* or the mechanism of heddle plus lifting device and *treadle*. In this book this term is used to mean the entire mechanism.

Shed: a temporary opening between warps, made during the weaving process for the insertion of weft. See also *natural shed*, *counter shed*, and *pattern shed*.

Shedding device: a tool for making an opening in a warp, for insertion of a weft.

Shed stick or rod: a stick used to save (hold open) a shed (opening) in the warp for weft insertion.

Shuttle: a tool used to insert wefts during weaving. Shuttle designs vary, but are commonly boat-shaped devices that enclose a spool with weft wound onto it.

Silk: a protein fiber obtained from an insect cocoon, most commonly the cocoons of the silk moth *Bombyx mori*.

Simple weave: a *weave structure* with one set of warp and weft. The term applies to basic weaves such as *plain weave*, *twill*, *satin*, and *gauze*. It also applies to a combination of weaves such as *damask*.

Single-heddle loom: a loom that has one heddle for opening the *counter-shed* for making the *ground weave*. A single heddle is normally used together with a *shed stick* that retains the *natural shed* opening in the warp, the weaver alternating between these two sheds when inserting weft. Other types of heddle, such as patterning heddles, may also be present on a "single-heddle" loom.

Slit tapestry: a type of tapestry where the wefts in adjacent color sections turn back at their terminal warps. The lack of a lateral connection between color sections creates the slits in the textile that give it its name.

Spindle: a stick used for spinning thread, sometimes weighted with a whorl made of wood, bone, stone or clay.

Spinning/spun thread: the process by which a fluffy short fiber such as *cotton* or *wool* is simultaneously spun and drawn out into a long thread.

Spool: see *shuttle*.

Supplementary warp: extra warp, in addition to ground warp. Supplementary-warp weaving is a method of producing patterns in which some warps, usually in a contrasting color to the rest of the warps, are allowed to "float" over several wefts (warp-float patterning) or pulled up as *pile* (*velvet*). Supplementary warps are generally continuous.

Supplementary weft: extra weft, in addition to ground weft, which is used for patterning. Supplementary weft usually has contrasting color to that of the ground weft. They can float over or wrapped around several warps to create patterns. Supplementary wefts can be *continuous* or *discontinuous*.

Sword/sword-beater: see *beater*.

Temple: a device in weaving, usually a stick with points at both ends, which is inserted into the woven cloth near its *selvedges*, providing tension across the warps and controlling the width of the finished textile.

Tenter: see *temple*.

Tapestry: *weft-faced* patterned cloth based on *plain weave* or *twill*, in which the ground wefts are discontinuous, usually in different colors, and woven back and forth within their own color sections for the pattern. There are several types, defined by the ways the boundaries between adjacent color sections are connected: *slit tapestry* (see also *kesi* and *kelim*); *dovetailed tapestry*, and *interlocked* (or double-interlocked) *tapestry*.

Taqueté: a weft-faced compound plain weave, with two sets of warps (called ground/foundation and inner warps) and two or more *complementary wefts* (usually in different colors). All wefts and the foundation warp interlace in plain weave. The inner warp separates the wefts: one color to appear on the front of the fabric for making the pattern, and the rest to remain on the back. Compare *samit*.

Textile: cloth that is produced by interlacing *yarns* in various ways. The definition used in this book includes textiles that are woven on a loom, and those that are made in other ways, such as knitting.

Thread: see *yarn*.

Thread count: the number of *warps* or *wefts* per unit length (generally 1 cm or 1 inch).

Throw: one insertion of weft through a *shed* opening, either using *shuttle* or *bobbin*.

Throw-shuttle: see *flying shuttle*.

Treadle: a bar pressed by the foot, used (for example) to raise a *heddle* or to depress a *shed stick* during weaving. Synonym: *pedal*.

Treadle loom: a loom equipped with *treadle*(s).

Triple cloth: a *compound weave* employing three sets of warps and wefts, each set interlaces to form its own weave and create a distinct layer.

Tubular warp: see *circular warp*.

Twill: a weave structure characterized by warp or weft floats and the diagonal bindings of the warp- and weft-interlacing. There are many different types of twill, such as diamond twill, herringbone, etc.

Twining: a textile structure in which one element is twisted together to enclose another element. In warp twining, for example, the warps are twisted together and enclose the wefts.

Velvet: a textile that is patterned using raised *pile* on the surface. The pile is made of loops of *supplementary warps* that are raised using metal rods inserted during weaving. The loops may later be cut to create plush or left uncut. Velvet piles can be of multiple heights depending on the size of the metal rods used.

Warp: the threads that make up the structure of a woven textile, with which *weft* is interlaced. On looms, warp threads run parallel and longitudinally.

Warp beam/axle: the beam to which the warp threads are attached or loop around. On some looms the warp threads are wound around this beam when weaving commences, being gradually unwound as the weaving progresses. On most horizontal looms the warp beam is the beam furthest from the weaver.

Warp-faced compound plain weave: a structure based on plain weave with multiple *complementary warps* (usually in different colors) and two sets of wefts (called ground/foundation and inner wefts). All warps and the foundation weft interlace in plain weave. The inner weft separates the warps: one color warp to appear on the front of the fabric for making the pattern, and the rest to remain on the back. Ancient Chinese *jin* textiles were woven using this structure.

Warp-faced weave: a woven surface on which the warp is the most prominent element.

Warp-float patterning: see *supplementary warp*.

Warp ikat: see *ikat*. In general, warp ikat is a *warp-faced weave*.

Warping: mounting a warp on a loom, before weaving.

Warp-patterned fabric: a woven cloth, generally a *warp-faced weave*, in which colored warps, for example in stripes, are the dominant patterning elements.

Warp-rib weave: a type of plain weave in which only the warp is visible, forming "ribs" of warp on the surface of the textile.

Warp selvedge: a *selvedge* made by turning back warps at the ends of a textile. This feature is found on some South American weavings and less common than weft selvedges.

Warp substitution: a technique in which warps, which are normally continuous throughout a textile, are replaced by warps of another color.

Warp twined weave, warp twining: see *twining*.

Weave structure: the interlacing of warps and wefts that repeats in regular manner to form a woven cloth. Weave structure can be broadly divided into *simple weave* (having one set of warp and weft) and *compound weave* (having more than one set of warp and weft).

Weaving: a process of making a textile by interlacing warps and wefts in a particular order with the aid of a device such as a loom.

Weft: the threads that make up the structure of a woven textile, with which *warp* is interlaced. In weaving, weft is inserted at right angles to the warp.

Weft-faced compound plain weave or twill: see *taqueté* and *samit*.

Weft-faced weave: a woven surface on which the weft is the most prominent element.

Weft-float patterning: see *supplementary-weft*.

Weft ikat: see *ikat*. In general, weft ikat is a *weft-faced weave*.

Weft-rib weave: a type of *plain weave* in which only the weft is visible, forming "ribs" of weft on the surface of the textile.

Weft-twined weave, weft twining: see *twining*.

Whorl: see *spindle*.

Wool: a protein fiber that is obtained from some animals, including sheep and some camelids, that has insulating properties and can be spun to make *yarn*.

Yarn: lengths of soft and flexible material; the basic raw material for making a textile.

Bibliography

Adrosko, Rita J. 1982. "The Invention of the Jacquard Mechanism." *Bulletin Du CIETA* 55–56: 89–117.

Afshar, Iraj. 1992. "*Zilu*." *Journal of Iranian Studies* 25 (1–2): 31–36.

Al'baum, Lazar. 1975. *Painting of Afrasiab*. Tashkent: FAN.

Andersson Strand, Eva. 2018. "Early Loom Types in Ancient Societies." In *First Textiles: The Beginnings of Textile Manufacture in Europe and the Mediterranean*, edited by Malgorzata Siennicka, Lorenz Rahmstorf, and Agata Ulanowksa, 17–29. Ancient Textiles Series. Oxford: Oxbow Books.

Andersson Strand, Eva, and Sara-Grace Heller. 2017. "Production and Distribution." In *Fashion in the Medieval Age (500–1450) A Cultural History of Dress and Fashion*, edited by Sara-Grace Heller, 29–52. Oxford: Oxbow Books.

Andersson Strand, Eva, Marie-Louise Nosch, and Linda Olofsson. 2015. "Experimental Testing of Bronze Age Textile Tools." In *Tools, Textiles and Contexts: Investigating Textile Production in the Aegean and Eastern Mediterranean Bronze Age*, edited by Eva Andersson Strand and Marie-Louise Nosch, 75–100. Oxford: Oxbow Books.

Baker, Chris. 2011. "Markets and Production in the City of Ayutthaya Before 1767: Translation and Analysis of Part of the Description of Ayutthaya." *Journal of the Siam Society* 99.

Barber, Elizabeth. 1992. *Prehistoric Textiles. The Development of Cloth in the Neolithic and Bronze Ages with Special Reference to the Aegean*. Princeton: Princeton University Press.

Barnes, Ruth. 1989. "The Bridewealth Cloth of Lamalera, Lembata." In *To Speak with Cloth: Studies in Indonesian Textiles*, edited by Mattiebelle Gittinger, 43–55. Los Angeles: UCLA Fowler Museum of Cultural History.

———. ed. 2010. *Five Centuries of Indonesian Textiles: The Mary Hunt Kahlenberg Collection*. Munich; New York: Delmonico Books/Prestel.

———. 2014. "Early Textiles from Timor." In *Textiles of Timor, Island in the Woven Sea*, edited by Roy W. Hamilton and Joanna Barrkman, 105–113. Fowler Museum Textile Series 13. Los Angeles: UCLA Fowler Museum of Cultural History.

———. 2018. "The Holmgren Spertus Collection of Indonesian Textiles in the Department of Indo-Pacific Art." *Arts of Asia* 48 (2): 68–80.

Bart, Bernhard. Forthcoming. "From the Khmer to the Minangkabau: A Technical Analysis of the Weaving Tools and Looms."

Baxter, M. J., H.E.M. Cool, and M. Anderson. 2010. "Statistical Analysis of Some Loom Weights from Pompeii: A Postscript." *Archeologia e Calcolatori* 21.

Beattie, May H. 1981. "A Note on *Zilu*." In *Flat-Woven Textiles*, edited by Cathy Cootner, 169–174. Washington D.C.: Textile Museum.

Belanová-Štolcová, Tereza, and Karina Grömer. 2010. "Loom-Weights, Spindles and Textiles. Textile Production in Central Europe from the Bronze Age to the Iron Age." In *North European Symposium for Archaeological Textiles X*, edited by Eva Andersson Strand, Margarita Gleba, Ulla Mannering, Cherine Munkholt, and Maj Ringgaard, 9–20. Ancient Textiles Series. Oxford: Oxbow Books.

Belenitsky, Alexsandr, Ilona Bentovich, and Oleg Bolshakov. 1973. *Medieval City of Central Asia*. Leningrad: Nauka.

Bichurin, Iakinf. 1950. *Collection of Information About the Peoples Who Lived in Central Asia in Ancient Times*. Moscow, Leningrad: AN SSSR Publ 2.

Bird, Junius Bouton. 1963. "Technology and Art in Peruvian Textiles." In *Technique and Personality*, edited by M. Mead and J. B. Bird, 45–78. New York: Museum of Primitive Art.

———. 1979. "New World Fabric Production and the Distribution of the Backstrap Loom." In *Looms and Their Products*, edited by Irene Emery, 115–126. Irene Emery Roundtable on Museum Textiles, 1977 Proceedings. Washington D.C.: Textile Museum.

Bird, Junius Bouton, and Milica Dimitrijevic Skinner. 1974. "The Technical Features of a Middle Horizon Tapestry Shirt from Peru." *Textile Museum Journal* 4 (1): 5–13.

Boland, Rita. 1977. "Weaving the Pinatikan, a Warp-Patterned Kain Bentenan from North Celebes." In *Studies in Textile History: In Memory of Harold B. Burnham*, edited by Veronica Gervers, 1–17. Ontario: Royal Ontario Museum.

Boudot, Eric. Forthcoming. "The Evolution of Figured Weaving in China Between the 3rd Century BCE and the 8th Century CE."

Boudot, Eric, and Christopher D. Buckley. 2015. *The Roots of Asian Weaving: The He Haiyan Collection of Textiles and Looms from Southwest China*. Oxford: Oxbow Books.

Boyce, Andrew, Barry Kemp, and Gillian Vogelsang-Eastwood. 2001. *The Ancient Textile Industry at Amarna. Excavation Memoir*. London: Egypt Exploration Society.

Broudy, Eric. 1979. *The Book of Looms. A History of the Handloom from Ancient Times to the Present*. Van Nostrand Reinhold.

Buckley, Christopher D. 2012. "Investigating Cultural Evolution Using Phylogenetic Analysis: The Origins and Descent of the Southeast Asian Tradition of Warp Ikat Weaving." *PLoS ONE* 7 (12): e52064.

———. 2017. "Looms, Weaving and the Austronesian Expansion." In *Spirits and Ships: Cultural Transfers in Early Monsoon Asia*. Singapore: ISEAS Publishing.

Buckley, Christopher D., and Eric Boudot. 2017. "The Evolution of an Ancient Technology." *Royal Society Open Science* 4 (5): 170208.

Cameron, Judith. 2004. *Cloth Production in the Prehistory of Southeast Asia*. PhD Dissertation, The Australian National University.

———. 2017. "A Prehistoric Maritime Silk Road: Merchants, Boats, Cloth and Jade." In *Beyond the Silk Roads: New Discourses on China's Role in East Asian Maritime History*, edited by R. J. Anthony and A. Schottenhammer, 25–43. East Asian Economic and Socio-Cultural Studies. East Asian Maritime History 14. Wiesbaden Harrassowitz Verlag.

Chapus, Georges Sully, and Berthe Dandouao. 1951–1952. "Les Anciennes Industries Malgaches." *Bulletin de l'Académie Malgache* 30.

Charlin, Jean-Claude. 2003. *Histoire de La Machine Jacquard, De l'Origine à Nos Jours*. Stäubli, Les Echets, Daniel Faurite.

Christie, Jan Wisseman. 1993. "Texts and Textiles in 'Medieval' Java." *Bulletin De l'Ecole Française d'Extrême-Orient* 80 (1): 181–211.

———. 2000. "Weaving and Dyeing in Early Java and Bali." In *Southeast Asian Archaeology 1998: Proceedings of the 7th International Conference of the European Association of Southeast Asia Archaeologists, Berlin, 31 August–4 September 1998*, edited by Wibke Lobo and Stefanie Reimann, 17–27. Hull; Berlin: Center for South-East Asian Studies, University of Hull; Ethnologisches Museum, Staatliche Museen zu Berlin Stiftung Preussischer Kulturbesitz.

Ciszuk, Martin, and Lena Hammarlund. 2008. "Roman Looms—A Study of Craftmanship and Technology in the Mons Claudianus Project." In *Vestidos, Textiles y Tintes. Estudios Sobre La Produccion de Bienes de Consumo En La Antiguedad. Actas Del II Symposium Internacional Sobre Textiles y Tintes Del*

Mediterraneo En El Mundo Antiguo (Atenas, 24 Al 26 de Noviembre, 2005), edited by C. Alfaro and L. Karali, 119–134. Textiles and Dyes in Antiquity. Valencia: Universitat de Valencia.

Clarke, Helen. 1986. *The Archaeology of Medieval England*. Oxford: Blackwell.

Coatsworth, Elizabeth, and Gale R. Owen-Crocker. 2017. "Textiles." In *Fashion in the Medieval Age (500-1450). A Cultural History of Dress and Fashion*, edited by Sara-Grace Heller, 11–28. Oxford: Oxbow Books.

Cohen, John. 2010. *Past Present Peru: Five Films*. Göttingen, Germany: Steidl.

Conklin, William J. 1978. "The Revolutionary Weaving Inventions of the Early Horizon Ñawpa Pacha." *Journal of Andean Archaeology* 16: 1–12.

———. 1979. "Moche Textile Structures." In *The Junius B. Bird Pre-Columbian Textile Conference, May 1973*, edited by Ann Rowe and Anne Shaeffer, 165–184. Washington D.C.: Textile Museum.

Costin, C. 1991. "Craft Specialization: Issues in Defining, Documenting, and Explaining the Organization of Production." *Archaeological Method and Theory*, 3: 1–56.

Cutler, Joanne. 2012. "Ariadne's Thread: The Adoption of Cretan Weaving Technology in the Wider Southern Aegean." In *KOSMOS. Jewellery, Adornment and Textiles in the Aegean Bronze Age. Proceedings of the 13th International Aegean Conference/13e Rencontre Égéenne Internationale, University of Copenhagen, Danish National Research Foundation's Centre for Textile Research, 21–26 April 2010. Aegaeum 33*, edited by Robert Laffineur and Marie-Louise Nosch, 145–154. Leuven; Liège: Peeters.

Del Freo, Maurizio, Marie-Louise Nosch, and Françoise Rougemont. 2010. "The Terminology of Textiles in the Linear B Tablets, Including Some Considerations on Linear a Logograms and Abbreviations." In *Textile Terminologies in the Ancient Near East and Mediterranean from the Third to the First Millennia BC*, edited by Cécile Michel and Marie-Louise Nosch, 338–373. Ancient Textiles Series. Oxford: Oxbow Books.

Desrosiers, Sophie. 1986. "An Interpretation of Technical Weaving Data Found in an Early 17th Century Chronicle." In *The Junius B. Bird Conference on Andean Textiles 1984*, edited by Ann P. Rowe, 219–241. Washington, D.C.: The Textile Museum.

———. 1997. "Lógicas Textiles et Lógicas Culturales en Los Andes." In *Saberes y Memorias en los Andes*, edited by T. Bouysse-Cassagne, 325–349. Paris; Lima: Institut Français d'Etudes Andines; Institut des Hautes Etudes sur l'Amérique Latine.

D'Harcourt, Raoul, 2002. *Textiles of Ancient Peru and Their Techniques*. Mineola, N. Y.: Dover Publications.

Diderot, Denis, and Jean Le Rond d' Alembert. 1765. *Encyclopédie Ou Dictionnaire Raisonné Des Sciences, Des Arts et Des Métiers*.

Donnan, Christopher B. and Sharon Donnan. 1979. "Moche textiles from Pacatnamu." In T*he Pacatnamu Papers. Volume II: The Moche Occupation*, edited by C. B. Donnan and G. A. Cock, 215–242. Los Angeles: UCLA Fowler Museum of Cultural History.

Doyon-Bernard, S. J. 1990. "From Twining to Triple Cloth: Experimentation and Innovation in Ancient Peruvian Weaving (ca. 5000–400 B.C.)." *American Antiquity* 55 (1): 68–87.

Dransart, Penny. 2007. "Mysteries of the Cloaked Body." In *The Nature and Culture of the Human Body: Lampeter Multidisciplinary Essays*, edited by P. Mitchell, 161–187. Trivium 37. University of Wales, Lampeter.

Emery, Irene. 1966. *The Primary Structures of Fabrics*. Washington D.C.: Textile Museum.

Flores Ochoa, J. A., K. MacQuarrie, and J. Portús. 1994. *Gold of the Andes: The Llamas, Alpacas, Vicuñas and Guanacos of South America*. Barcelona: Patthey & Sons.

Fraser-Lu, Sylvia. 1988. *Handwoven Textiles of South-East Asia*. Singapore; Oxford: Oxford University Press.

Garcilaso de la Vega, Inca. 1609, 1617. *Royal Commentaries of the Incas and General History of Peru*. Translated and edited by Harold V. Livermore, 2 Vols., 1966. Austin and London: University of Texas Press.

Gavin, Traude. 1996. *The Women's Warpath: Iban Ritual Fabrics from Borneo*. Los Angeles: UCLA Fowler Museum of Cultural History.

Gittinger, Mattiebelle. 1979. "An Introduction to the Body-Tension Looms and Simple Frame Looms of Southeast Asia." In *Looms and Their Products: Irene Emery Roundtable on Museum Textiles, 1977 Proceedings*, edited by Irene Emery, 54–68. Washington D.C.: Textile Museum.

Gleba, Margarita. 2008. *Textile Production in Pre-Roman Italy*. Ancient Textiles Series. Oxford: Oxbow Books.

Guaman Poma de Ayala, Felipe. 1615–1616. *El Primer Nueva Corónica y Buen Gobierno*. København: Det Kongelige Bibliotek, GKS 2232 4°.

Hamilton, Roy W., ed. 1994. *Gift of the Cotton Maiden: Textiles of Flores and the Solor Islands*. Los Angeles: UCLA Fowler Museum of Cultural History.

———. 1998. *From the Rainbow's Varied Hue: Textiles from the Southern Philippines*. Los Angeles: UCLA Fowler Museum of Cultural History.

Hauser-Schäublin, Brigitta, Marie-Louise Nabholz-Kartaschoff, and Urs Ramseyer. 1991. *Balinese Textiles*. London: British Museum Press.

Heidmann, P. 1937. *La Revue de Madagascar*. Tananarive: Imprimerie Officielle: 17.

Hoffmann, Marta. 1964. *The Warp-Weighted Loom. Studies in the History and Technology of an Ancient Implement*. Studia Norvegica. Oslo: Universitetsforlaget.

Howard, Michael. 2008. "Supplementary Warp Patterned Textiles of the Cham in Vietnam." *Textile Society of American Symposium Proceedings*.

Jasper, J. E., and Mas Pirngadie. 1912. *De Inlandsche Kunstnijverheid in Nederlandsch Indië II: De Weefkunst*. The Hague: Mouton & Co.

Kriger, Colleen E. 2006. *Cloth in West African History*. Oxford: Alta Mira Press.

Loir, J. 1926. *Théorie Du Tissage Des Étoffes de Soie*. Vol. 3. Lyon; Paris: Joannès Desvigne et Cie, Successeurs; Librairie Scientifique Desforges.

Mannering, Ulla. 2011. "Early Iron Age Craftsmanship from a Costume Perspective." In *Arkæologi I Slesvig. Archäologie in Schleswig. Det 61. Internationale Sachsen Symposion 2010, Haderslev, Danmark*, edited by Linda Boye, 85–94. Neumünster: Wachholtz Verlag.

Maxwell, Robyn. 2003. *Textiles of Southeast Asia: Tradition, Trade and Transformation*. Singapore: Periplus.

McIntosh, Linda S. 2008. *Weaving Paradise: Southeast Asian Textiles and Their Creators*. Bangkok: James H. W. Thompson Foundation.

———. 2009. *Textiles of the Phu Thai of Laos*. PhD Dissertation, Simon Fraser University.

———. 2014. *Carving a Community: The Katu People*. Luang Prabang. Laos: Traditional Arts and Ethnology Centre.

Mokdad, Ulrikke. 2014. "Gobeliner Fra Serbien-Nu Med Dansk Islat." *RAPPORTER Fra Tekstilernes Verden* 3.

Mollet, Louis. 1951–1952. "Métier Betsimisaraka à Deux Rangs de Lisses." *Bulletin de l'Académie Malgache* 30.

Moshkova, Valentina. 1970. *Carpets of Central Asia of the Late XIX – Early XX Centuries*. Tashkent: FAN.

Narshahi, Muhammad. 1897. *History of Bukhara*. Tashkent.

Østergård, Else. 2003. *Woven into the Earth. Textiles from Norse Greenland*. Aarhus: Aarhus Universitetsforlag.

Pastor-Roces, Marian, Dick Baldovino, and Wig Tysmans. 1991. *Sinaunang Habi: Philippine Ancestral Weave*. The Nikki Coseteng Filipiniana Series. Quezon City, Philippines: N. Coseteng.

Phipps, Elena. 1982. *Discontinuous Warp and Weft in Andean Weaving*. MA thesis, Columbia University.

———. 2013. *The Peruvian Four-Selvaged Cloth: Ancient Threads/New Directions*. Los Angeles: UCLA Fowler Museum of Cultural History.

Phipps, Elena. 2004. *The Colonial Andes: Tapestries and Silverwork, 1530-1830*. New York: Metropolitan Museum of Art.

Picton, John, and John Mack. 1989. *African Textiles*. Second, revised edition. London: British Museum Publications.

Pugachenkova, G. A., E. V. Rtveladze, and K. Kato. 1991. *Antiquities of Southern Uzbekistan*. Tashkent: Soka University.

Rawson, Jessica. 1983. *The Chinese Bronzes of Yunnan*. London: Sidgwick & Jackson.

Roth, H. Ling. 1934. *Studies in Primitive Looms*. 2nd ed. Halifax: King & Sons.

Rowe, Ann P. 1977. *Warp-Patterned Weaves of the Andes*. Washington D.C.: The Textile Museum.

Scherrer, Guy. 1993. "Quelques Aspects Oubliés Du Tissage de La Soie Au XVIIIe Siècle." *Bulletin Du CIETA* 71: 90–95.

———. 2006. "Les Premiers Cartons de Jacquard." *La Revue Du Musée Des Arts et Métiers* 45: 55–61.

———. 2012. "Soierie: Autour de l'Outil de Travail Au XVIIIᵉ Siècle." In *Lyon Au XVIIIᵉ, Un Siècle Surprenant!, Catalogue d'Exposition, Novembre 2012*, 100–107. Musées Gadagne, Lyon / Somogy éditions d'Art.

Spieth, Jacob. 1906. *Die Ewe-Stämme: Material Zur Kunde Des Ewe-Volkes in Deutch-Togo*. Berlin: Dietrich Reimer.

Strelow, Renata. 1996. *Gewebe Textiles*. Staatliche's Museum zu Berlin.

Stübel, Hans. 1937. *Die Li Stämme der Insel Hainan*. Berlin: Klinkhardt and Biermann, Verlag.

Sukhareva, Olga. 1962. *Late Feudal City of Bukhara*. Tashkent.

Summerfield, Anne, John Summerfield, and Taufik Abdullah. 1999. *Walk in Splendor: Ceremonial Dress and the Minangkabau*. Los Angeles: UCLA Fowler Museum of Cultural History.

Sun, Guoping, and Weijin Huang. 2004. "The Emergence of the Tianluoshan Site, Yuyao City, Zhejiang Province." *Zhongguo Wenwu Bao* 6 (1).

Thompson, Jon, and Hero Granger-Taylor. 1995–1996. "The Persian *Zilu* Loom of Meybod." *Bulletin Du CIETA* 73: 27–53.

Van Hout, Itie. Forthcoming. "An 'Enchanting' Technique: Twill Weaving in East Kalimantan, Indonesia."

Vanstan, Ina. 1979. "Did Inca Weavers Use An Upright Loom?" In *The Junius B. Bird Pre-Columbian Textile Conference, May 1973*, edited by Ann Rowe and Anne Shaeffer, 233–239. Washington, D.C.: Textile Museum.

Veth, Pieter Johannes. 1882. *Midden-Sumatra: Reizen Und Onderzoekingen Der Sumatra-Expeditie, Uitgerust Door Het Aardrijk-Skundig Genootschap, 1877–1879*. Leiden: E. J. Brill.

Vogelsang-Eastwood, Gillian. 1988. "*Zilu* Carpets from Iran." *Studia Iranica* 18: 225–240.

———. 1992. *The Production of Linen in Pharaonic Egypt*. Leiden: Textile Research Centre.

Wild, John Peter. 1970. *Textile Manufacture in the Northern Roman Provinces*. Cambridge: Cambridge University Press.

Yin, Shaoting, Xuehui He, Yu Luo, and Qiu Zhong. 1999. *The Material Culture of Yunnan: Spinning and Weaving*. Kunming: Yunnan Education Publishing House.

Yoshimoto, Shinobu. 1990. "Typological Studies of Indonesian Handlooms: (1) Types and Distribution." *Bulletin of the National Museum of Ethnology* 15 (1): 1–114.

Zhang, Jianping, Houyuan Lu, Guoping Sun, Rowan Flad, Naiqin Wu, Xiujia Huan, Keyang He, and Yonglei Wang. 2016. "Phytoliths Reveal the Earliest Fine Reedy Textile in China at the Tianluoshan Site." *Scientific Reports* 6 (January): 18664. doi: 10.1038/srep18664.

Zhao, Feng. 1992. "Restoration of Liangzhu Loom." *Southeast Culture* 2: 108–111.

———. 1994. "A Study on the Vertical Treadle Loom." *Studies in the History of Natural Sciences* 2: 145–154.

———. 1996. "Type and Spread of Backstrap Treadle Loom." *Journal of Zhejiang Institute of Silk Textiles* 5: 18–25.

———. 1997. "Reconstruction of Axle-Treadle Loom in Han Dynasty." *Journal of China Textile University* 4 (English Edition): 60–65.

———. 2012. *Studies of Textiles and Weaving Technology (7th Century BCE–3rd Century BCE)*. Shanghai: Shanghai Chinese Classics Publishing House.

———. 2016. *Silks from the Silk Road: Origin, Transmission and Exchange*. Hangzhou: Zhejiang University Press.

Zhao, Feng, Yi Wang, Qun Luo, Bo Long, Baichun Zhang, Yingchong Xia, Tao Xie, Shungqing Wu, and Lin Xiao. 2017. "The Earliest Evidence of Pattern Looms: Han Dynasty Tomb Models from Chengdu, China." *Antiquity* 91: 360–374.

Zhejiang Provincial Institute of Cultural Relics and Archaeology Research. 1988. "Briefing on the Excavation of Liangzhu Tomb in Fanshan, Yuhang, Zhejiang." *Cultural Relics* 1: 1–31.

Zhou, Daguan. 2007. *Customs of Cambodia.* Translated by Peter Harris. Chiang Mai: Silkworm Books.

图书在版编目（CIP）数据

神机妙算：世界织机与织造艺术 ＝ A World of Looms: Weaving Technology and Textile Arts：英文 / 赵丰，（美）桑德拉，（英）白克利主编. —杭州：浙江大学出版社，2019.10（2020.3 重印）

ISBN 978-7-308-19182-1

Ⅰ．①神… Ⅱ．①赵… ②桑… ③白… Ⅲ．①纺织机械—介绍—世界—英文 Ⅳ．① TS103

中国版本图书馆 CIP 数据核字（2019）第 107309 号

神机妙算：世界织机与织造艺术
赵　丰　[美]桑德拉　[英]白克利　主编

策　　划	张　琛　包灵灵
责任编辑	黄静芬
责任校对	祁　潇
封面设计	周　灵
出版发行	浙江大学出版社
	（杭州市天目山路 148 号　邮政编码 310007）
	（网址：http://www.zjupress.com）
排　　版	周　灵
印　　刷	浙江印刷集团有限公司
开　　本	889mm×1194mm　1/16
印　　张	19.5
字　　数	854 千
版 印 次	2019 年 10 月第 1 版　2020 年 3 月第 2 次印刷
书　　号	ISBN 978-7-308-19182-1
定　　价	360.00 元

审图号：GS（2019）3172号
版权所有　翻印必究　　印装差错　负责调换
浙江大学出版社市场运营中心联系方式：0571-88925591；http://zjdxcbs.tmall.com